First You Build a Cloud

K. C. COLE

First You Build a Cloud

And Other Reflections on Physics as a Way of Life

A HARVEST ORIGINAL
HARCOURT BRACE & COMPANY
San Diego New York London

Library of Congress Cataloging-in-Publication Data
Cole, K. C.
First you build a cloud: and other reflections on physics
as a way of life/by K. C. Cole.—Rev. ed.
p. cm.
ISBN 0-15-600646-4
Originally published: Sympathetic vibrations.
1st ed. New York: W. Morrow, ©1985
Includes index.
1. Physics. I. Cole, K. C. Sympathetic vibrations. II. Title.
QC21.2.C62 1999
530—dc21 98-47050

Designed by Linda Lockowitz
Text set in Centaur
Printed in the United States of America
First Harvest edition 1999
A C E F D B

Newton himself, as well as those... who attacked him... would have all alike been amazed at the more recent contention that natural science has nothing to do with "values," that it can and should itself remain "value-free," and that those seeking a direction for human life have nothing to learn from our best knowledge of the nature of things. Even a little science... is a thing of infinite promise for human values.

—JOHN HERMAN RANDALL, JR., *Newton's Philosophy of Nature*

Physical science used to be called natural philosophy, but unfortunately physics is no longer taught as a course in philosophy. This situation is surprising, because most of the physicists that I know talk with me and with each other with full awareness that the way in which we think of the physical world profoundly shapes the way we think of the human and ethical worlds. For them physics is a part of culture and of philosophy. In the chapters of this book, K. C. Cole has been able to talk to nonphysicists in the thoughtful way that many physicists talk to each other.

—FRANK OPPENHEIMER

Founder of The Exploratorium, a museum of science and perception in San Francisco

CONTENTS

For Bob Miller

THIS NEW VERSION of the book formerly known as *Sympathetic Vibrations: Reflections on Physics as a Way of Life* has been substantially revised. Much new material has been added, and many sections have been dramatically honed or deleted. In addition, the material has been reorganized into three thematic parts.

The introduction, Living in Outer Space, lays out the philosophical framework for the book, and explains certain biases that guide the exposition and tone.

The first part, The Art of Knowing, explores the many approaches physicists take in their attempts to see the unseeable and understand a material universe that may well turn out to be beyond human ken. The second, Movers and Shakers, deals with the forces and pseudoforces that make the world go round, as well as the foundations of modern physics: quantum mechanics and Einstein's special and general theories of relativity. The third, Threads and Knots, follows some of the repeating patterns that weave throughout the physical universe. Some, like sympathetic vibrations, are rhythms that seem to pop up everywhere one looks in nature; others, like the relationship between cause and effect, leave physicists facing knotty problems they're still trying hard to unravel.

ACKNOWLEDGMENTS

It's a rare treat for an author to get a second crack at an already published manuscript, so special thanks to Jane Isay, my editor at Harcourt Brace, for the chance to turn the book formerly known as *Sympathetic Vibrations* into *First You Build a Cloud.* I like the book much better now, and I hope she does. Thanks also to Jane's assistant, Lorie Stoopack, for good editorial advice and unfailing good humor when I needed it most.

I continue to be grateful to all the editors down the line who encouraged me to explore the connections between physics and philosophy in popular publications: Phyllis Theroux, who first suggested that I try writing about the unlikely subject of physics as a way of life in four essays for the *Washington Post;* Nancy Newhouse of *The New York Times,* for encouraging me to pursue the same subject in a series of seven "Hers" columns; Leon Jaroff, founding editor of *Discover* who gave me a column in his magazine; and Evelyn Renold, formerly at *Lears,* who let me prove the appeal of physics to "women who weren't born yesterday."

I owe much of this book to my friend the physicist, Frank Oppenheimer, who inspired and encouraged me, and told me what to read and who to talk to. Among those wonderful mentors and teachers were Philip Morrison and Victor Weisskopf. Both Weisskopf and Oppenheimer criticized the first version of this book with great (and ruthless) care—for which I am eternally indebted. Thanks also to the staff of The Exploratorium—especially the "dinosaurs"—a wellspring of inspiration for twenty-five years. And to Nancy Rodgers, for her illuminating photographs.

Finally, I wish to acknowledge the role of Bob Miller, to whom this book is dedicated. Artist, natural philosopher, and great family friend, Bob has taught us about living in outer space and the secret lives of shadows, the universe inside a pinhole and the awesome weight of clouds. Most of all, thanks to Bob for his unflagging vigilance against the pervasive threat he calls "hardening of the categories."

First You Build a Cloud

INTRODUCTION

Living in Outer Space

Books on physics are full of complicated mathematical formulae. But thought and ideas, not formulae, are the beginning of every physical theory.

—ALBERT EINSTEIN AND LEOPOLD INFELD, *The Evolution of Physics*

The discoveries of science, the new rooms in this great house, have changed the way people think of things outside walls.... It is my thesis that [these discoveries] do provide us with valid and relevant and greatly needed analogies to human problems lying outside the present domain of science or its present borderlands.

—J. ROBERT OPPENHEIMER, *Science and the Common Understanding*

POPULAR SCIENCE writers are forever proclaiming the profound importance of such matters as the ultimate fate of the universe, or the events that took place during the first trillionth of a second of time. We write as if people were poised on the edges of their proverbial seats, anxiously waiting to learn whether or not the proton will decay (in 10^{32} years!) or whether there is mass tucked away mysteriously inside neutrinos.[1] Breathlessly, we keep them up to date on such esoterica as the search for the "top

[1] 10^{32} is the number 1 followed by thirty-two zeros. To give you an idea of how large that is, consider that a billion is the number 1 followed by only nine zeros. Physicists think the universe is about 15 billion years old. Waiting for a proton to decay is like waiting for Godot.

quark" or the "cosmological constant." Sometimes I wonder how many readers struggling to get their socks on in the morning doubt whether exploring these far-out corners of the universe is necessarily so important. It is surely not obviously useful.

But the juiciest mysteries of the universe at large and the innards of the atom are closer to home than we think. Science fiction fantasies describe travels into "outer space" as if it were some strange and exotic landscape, the site of galaxies far, far away and the possible home of weird creatures. Yet our watery world rolls around all day long in outer space, and at least some of those weird creatures are us—made of flesh and bone and, ultimately, quarks. Outer space is not alien territory. Neither is inner space. It's where we live.

Perhaps more compelling, the ideas of science have seeped into almost every aspect of our language and culture. We speak of people who attract and repel each other (like magnets), of the force of habit, of cause and effect, of disorder and quantum leaps, most of all, of space and time. Our language is liberally sprinkled with the metaphors of science—and the language of science is inescapably infused with images from everyday life.

Most scientists are rightly suspicious of attempts to apply their sharply defined ideas to the murky realm of human affairs. Yet just as many scientists think that science and the humanities have suffered a painful and unnatural separation. Far more than a collection of facts, science is a body of ideas that forms the cultural context through which we view the world. It influences, and is influenced by, almost everything else around us, from the role of religion to the status of slaves. Science started out as "natural philosophy." When the so-called scientific revolution of the seventeenth century produced such signal works as Kepler's *Harmonics of the World* and Galileo's *Starry Messenger*, the new discoveries were considered part of what was then called the "new philosophy." This, too, should not surprise us. Both philosopher and physicist are concerned with the causes of things—with the questions, Why are things the way they are? and Why do they behave the way they do?

Lately scientists have largely lost their status as philosophers. Many people think of science primarily in terms of technology. Yet history shows that science has long been a shaper of human thought. As the late physicist Max Born put it, "It has been said that the metaphysics of any period is the offspring of the physics of the preceding period. If this is true, it puts us physicists under the obligation to explain our ideas in a not-too-technical language."

When Born was writing in the early part of this century, many of his colleagues were also trying to untangle and explain the philosophical implications of their own scientific revolution—the revolution that rode on the wave of relativity and quantum theory and completely changed the way we think about everything from time and space to energy and matter.

Probably the best popular book on Einstein's theory, *The Universe and Dr. Einstein,* by Lincoln Barnett, was published in 1948, when Einstein was still alive and the ramifications of the "new physics" were still being unraveled. (Of course, they are still being unraveled, but there seemed to be more of a sense then that these subjects were somewhat wet behind the ears, and needed exploring and explaining.) In his introduction, Barnett explained his reasons for writing:

> Today most newspaper readers know vaguely that Einstein had something to do with the atomic bomb; beyond that his name is simply a synonym for the abstruse.... Many a college graduate still thinks of Einstein as a kind of mathematical surrealist rather than as the discoverer of certain cosmic laws of immense importance in [our] slow struggle to understand

> physical reality. He may not realize that Relativity, over and above its scientific import, comprises a major philosophical system which augments and illumines the reflections of the great epistemologists—Locke, Berkeley, and Hume.

The idea that science is inseparable from philosophy is a theme that pervades this book.

The other obvious prejudice is a preference for what physicist Victor Weisskopf always called "the old stuff."

In short, this book is not about black holes. Or quarks, or antimatter, or high-temperature superconductors, or the fate of the universe. It touches on these things and they are interesting enough in their own right. But concentrating on them tends to promote the feeling that science is something outside our everyday experience. And science is no more "inaccessible" than looking out the window and wondering why a tree branches in a certain way or why (to ask an old but still wise question) the sky is blue.[2]

I like old stuff because it takes no more knowledge or equipment than that available to, say, someone taking a flight on a commercial airliner. A dozen times a year, I find myself a passenger on one of these "space" trips, forty thousand feet in the air and awed, as always, by the oddity of jet-age flight: my five-hundred-thousand-pound airplane held aloft by unseen forces, no strings attached. The earth below is no longer a world but a spinning sapphire floating in space, veiled by an improbable icing

[2]Many years ago, my son asked an interesting variation of this question: What color is air? After much thought, we finally came up with the answer: blue. Air is blue for the same reason that the sky (which is made of air) is blue—namely, because clusters of air molecules scatter blue light more than all the other colors in sunlight. Air doesn't look as blue as the sky only because there isn't as much of it in a small space. (You might say that air is very, very, very, very light blue.)

of clouds. From this perspective, you can clearly see its curvature, be amazed by its smallness—a soft, sunlit spot of blue in a great dark void.

I can look up and see that I am practically bumping my head on the ceiling of space; another forty thousand feet or so and I would be in the daytime darkness—a rocketwoman. I look down into a bubbling cauldron of clouds, pierced by an occasional unfriendly mountain peak. We seem to be skimming the surface of an angry sea, and I think it unbelievable that delicate life-forms can exist down there—much less build enormous, noisy, metallic grasshoppers like the one I am flying in to pick them up and hop them from one port to another.

Many years ago, traveling with my six-year-old son, I oohed at the great white splotches of salt on the surface below, like so much spilt milk—the salt of the earth. I aahed at the jagged white spine of the Rockies that slashed the continent down its middle, the twisted river veins that brought life to the blue-blooded planet, the valleys etched so deeply that you could almost feel the pain it took to carve them out of the living planet's face. My son watched the movie. Unable to contain myself, I urged him to look down. He seemed not at all surprised, only vaguely interested. Finally I asked him what he thought was holding the huge plane and its hundreds of people up in the sky. As if he were talking to a child himself, he answered, "Air, of course."

Of course.

At that moment, the stewardess came by and asked me to lower my window shade. Other people, she said, wanted to watch the movie. Poor old earth, I thought. Pockmarked by comets, wrinkled and worn by wind and rain, patches of new green growth poking up everywhere, incredibly, in the ashes of the old—and no one to admire your dignified beauty. Talk about taking things for granted!

Artist Bob Miller likes to ask the following "science" question: How would you hold a hundred tons of water in thin air with no visible means of support? Answer: Build a cloud.

Thoreau knew that nature was a "wizard," but we seem to have forgotten. "The evolution of a lifeless planet eventually culminates in green leaves," writes naturalist Loren Eiseley. "The altered and oxygenated air hanging above the continents presently invites the rise of animal apparitions compounded of formerly inert clay. Only after long observation does the sophisticated eye succeed in labeling these events as natural rather than miraculous."

As the great British experimentalist Michael Faraday put it, "Nothing is too wonderful to be true." MIT theorist Philip Morrison, in a book with Faraday's words as the title, elaborates:

> Faraday was cogent. Nothing is too wonderful to be true: not an earth ball that pulls centrally on all alike, not the many invisible moons of Jupiter, not the single selfsame form of nerve impulses, whether the internal message they carry is that of sight or sound, not the count of sun-like stars in the visible cosmos, large enough to allot about ten billion suns to each human being alive, not even the slow enduring drift of India out of the southern ice to collide with Asia, there to raise up the mighty Himalaya.

And it's all old stuff, every wonderful bit of it. However, there's an even more fundamental reason for my concentration on everyday "old" stuff: What's fascinating about science at the forefront is only an embellishment of what's fascinating about science in everyday life. Gravity is at the bottom of black holes. In fact, a black hole is just a way of looking at what gravity would do if pushed (or pulled) to extremes. And gravity itself is a deep (and unsolved) puzzle.

In the same way, the behavior of supercold materials seems

sometimes supernatural: The so-called superfluids that exist only around absolute zero (minus 459 degrees Fahrenheit) can flow up and out of bottles, down through the bottoms of ceramic containers; superconducting materials can carry an electric current forever without offering the slightest resistance. At the hottest end of the temperature scale, matter also acts strangely. Atoms fall apart, forming superhot, electrically charged plasmas that ignite fusion fires like those in the center of the sun. To tame these elusive, too-hot-to-touch gases, physicists have constructed huge magnetic bottles, but a plasma is a slippery thing and hard to contain. Plasmas are the stuff of stars. At hotter temperatures still, physicists hope to create the so-called quark-gluon plasma, the primordial cosmic soup from which all else in the universe eventually condensed.

Yet all these exotic forms of matter are nothing more than extensions of the familiar spectrum of states of matter—from solid to liquid to gas. And it is impossible to appreciate either superconductivity or primordial soup without understanding the transformation of water to ice and steam.

Some people say that subjects like gravity or the states of matter are too fundamental to be interesting. People today are too sophisticated. Yet it's amazing how easy it is to be clueless even in this most technical of modern worlds. "Most of us are in daily contact with at least as much that we do not understand as were the Greeks or early Babylonians," my friend the physicist liked to say. "Yet we have learned not to ask questions about how the power steering on our cars works or how polio vaccine is made or what is involved in the freezing of orange juice. We end up in the paradoxical situation in which one of the effects of science is to dampen curiosity."

If simple science is uninteresting, it may be only because we have been made ashamed of asking all those simple, "obvious" questions. We still don't know for sure where the moon came from, or how life arose on earth, or why protons weigh more than electrons. We don't know why people respond to music or whether it will rain tomorrow in Minneapolis. We don't know the nature of evil or of the force that binds quarks together. "The genius of men like Newton and Einstein," writes Jacob Bronowski in *The Ascent of Man,* "lies in that: They ask transparent, innocent questions which turn out to have catastrophic answers. Einstein was a man who could ask immensely simple questions."

Clearly, this book is about more than the practical and philosophical fruits of science; it also celebrates what my friend the physicist liked to call "the sentimental fruits of science."

"Science is useful not only in a practical way," he said, "but also in that it determines how we think and feel. Religions have always embodied a view of nature. Even the Bible begins with an account of cosmology. Today such thoughts about nature come primarily from science. They are as imaginative and as fantastic as ever. But today people ascribe a very limited role to science. They continue to talk of the arts and music as culture, but neglect the fact that our view of ourselves and our perception of what our world is like are equally and vitally a part of culture."

Almost thirty years ago, I sat in the musty lobby of a small hotel in the Soviet industrial city of Kharkov, having just returned from a visit to a collective farm. A small crowd of Americans and Russians squirmed together in the tiny uncomfortable room to watch on a scratchy black-and-white TV as two American astronauts walked on the moon. The image on the screen was a barely discernible blur, yet it was quite clear that Russians and Americans alike were deeply impressed by these first tentative extraterrestrial steps. It was, for all of us, a profoundly emotional experience—like seeing the moons of Jupiter for the first time through a pair of good binoculars, or the rings of Saturn.

Even before NASA propelled us out of this world, the invention of the telescope "dramatically changed our *cultural* history," according to Isaac Asimov (emphasis his): "When Galileo looked at the moon with a telescope and saw mountains, craters and 'seas,' that was the final piece of evidence in favor of a plurality of worlds. Earth was not the only object on which life could conceivably exist." The telescope so expanded our view of the universe that "the great man-centered drama of sin-and-redemption, constructed in earlier times, looked puny against the new universe." It was the telescope that finally and firmly plucked people from their center-stage spot in the cosmos. Although some people had obviously figured out as far back as the third century B.C. that the earth orbited the sun (and not vice versa), this wasn't incorporated into the popular culture until well after Copernicus, in the sixteenth century. Conventional wisdom put the earth at the center of things; the universe was here for us. Imagine what that meant for people's sense of destiny, personal responsibility, and awe.

In some ways, however, the things we've learned through science have shown us that the earth is more central than ever—that life is even more precious because of its improbability. The stuff we're made of was forged in exploding stars. We bask in the light of a second-generation sun, our planet's composition a consequence of a slight contamination by foreign elements in the original hydrogen gas cloud that formed the solar system. In a very real sense, our rocky home is the sediment that sank to the bottom (really the center, but then "down" is always toward the center of the earth) when the lighter elements were blown or boiled away. Animals arose on land only after early bacteria accidentally polluted their environment with a "poison" called oxygen. This knowledge has not necessarily made us more or less humble, but it has certainly, said my friend the physicist, "changed the nature of our humility."

A closer look at the seemingly static cosmos also shows that it is dizzy with change; even the stars use up their resources, die, and are born again. But the sometimes violent births and deaths of stars, the continual evolution of the universe, was unknown only a few centuries ago. Until the time of Galileo, it was simply assumed that the stars we see today are the same stars that existed at the beginning of time, the same stars that would exist forever. •

At the opposite end of the size scale, the twentieth-century development of quantum theory shattered the notion that atoms behave like billiard balls, and that everything they (and therefore we) do is predetermined. A great deal of uncertainty lies at the heart of atoms; the meaning of a seemingly simple idea like cause and effect turns out to be immensely rich and complicated. But the result is that things today look far more flexible than they did in Newton's clockwork, preset universe.

Charles Darwin's formidable (and in some corners still forbidden) fruit was the knowledge that species, like stars, can change. The forms of life that inhabit the earth are not immutable. We, like the universe, *evolve.* Strict biblical creationists wouldn't be creating such a stir about Darwin if the question of where we came from and what our ancestors looked like wasn't an issue that itself stirred deeply within our souls.

Even our view of strictly physical forces such as gravity can have a profound effect on the way we view ourselves. Newton blew up a storm in the prevailing cultural winds of the seventeenth century with his universal theory of gravitation not because he "discovered" gravity (everyone knew things fell toward the earth) but because he discovered that gravity was universal. Before him, it was assumed that the laws of nature on earth were fundamentally different from those in the heavens. Newton showed that the fall of the apple and the orbit of the moon were controlled by the same forces.

In this sense, the need to go to the moon or smash atoms is on a par with the need to have natural history museums: Science provides a handle on who we are and how we fit into the scheme of things. Understanding our place in the sun requires an understanding of the sun's place in the solar system, the cycles of the sky, the nature of the elements, and the improbabilities of life. If what we learn leaves us a little stunned by our limitations and potentials, so be it. Science gives us a sense of scale and a sense of limits, an appreciation for perspective and a tolerance for ambiguity.

The best summary I ever read of this sentiment came from Robert R. Wilson, the sculptor and physicist who built the giant atom smasher at the Fermi National Accelerator Laboratory, near Chicago. He expressed it in response to the continual questions of a senator who demanded to know what probing protons had to do with the national defense:

"Is there anything connected in the hopes of this accelerator that in any way involves the security of the country?" asked the senator.

"No, sir, I do not believe so," responded Dr. Wilson.

"It has no value in that respect?" asked the senator again.

"It only has to do with the respect with which we regard one another, the dignity of people, our love of culture. It has to do with these things: Are we good painters, good sculptors, great poets? I mean all the things that we really venerate and honor in our country and are patriotic about.

"In that sense, this new knowledge has all to do with honor and country, but it has nothing to do directly with defending our country—except to help make it worth defending."

PART I:

The Art of Knowing

In our endeavor to understand reality we are somewhat like a man trying to understand the mechanism of a closed watch. He sees the face and the moving hands, even hears its ticking, but he has no way of opening the case. If he is ingenious he may form some picture of a mechanism which could be responsible for all the things he observes, but he may never be quite sure his picture is the only one which could explain his observations.

—Einstein and Infeld, *The Evolution of Physics*

CHAPTER ONE

Science as Metaphor

> At the leading edge of experience in philosophy, science and feeling there is inevitably a groping for language to translate the insecure novelty of noticing and understanding into a precision of meaning and imagery.
>
> —Frank Oppenheimer

Oppenheimer wrote these words in the introduction to a series of readings at The Exploratorium on "The Language of Poetry and Science." Poetry and science? Not so strange when you consider that Niels Bohr himself once wrote, "When it comes to atoms, language can be used only as in poetry. The poet, too, is not nearly so concerned with describing facts as with creating images."

Science, after all, involves looking mostly at things we can never see. Not only quarks and quasars but also light "waves" and charged "particles"; magnetic "fields" and gravitational "forces"; quantum "jumps" and electron "orbits." In fact, none of these phenomena is literally what we say it is. Light waves do not undulate through empty space in the same way as water waves ripple over a still pond; a field is not like a hay meadow, but rather a mathematical description of the strength and direction of a force; an atom does not literally leap from one quantum state to another; and electrons do not really travel around the atomic nucleus in circles any more than love produces literal heartaches. The words we use are metaphors, models fashioned from familiar ingredients and nurtured with imagination. "When a physicist says 'an electron is like a particle,'" writes physics professor Douglas Giancoli "he is making a metaphorical comparison, like the poet who says, 'love is like a rose.' In both images a concrete object, a rose or a particle, is used to illuminate an abstract idea, love or electron."

Over the centuries the metaphors of science have taken a multitude of forms. Recently, physicists struggling to understand new evidence for a repulsive force in the universe could be heard tossing around terms like "quintessence," "X matter," "smooth stuff," and "funny energy." The more mysterious the emerging landscape, the further they must reach for appropriate imagery to describe it. But there's nothing necessarily odder about this language than the terms scientists have always used to pin down the ineffable.

Here's Francis Bacon's seventeenth-century description of heat: "Heat is a motion of expansion, not uniformly of the whole body together, but in the smaller parts of it, and at the same time checked, repelled, and beaten back, so that the body acquires a motion alternative, perpetually quivering, striving, and irritated by repercussion, whence spring the fury of fire and heat."

And Isaac Newton's account of what we now call chemical reactions: "And now we might add something concerning a most subtle spirit which pervades and lies hid in all gross bodies, by the force and action of which spirit the particles of bodies attract one another at near distances and cohere, if contiguous... and there may be others which reach to so small distances as hitherto escape observations... and electric bodies operate to greater distances, as well repelling as attracting the neighboring corpuscles; and light is emitted, reflected, refracted, inflected, and heats bodies; and all sensation is excited and... propagated along the solid filaments of the nerves."

And Hans Christian Oersted's early-nineteenth-century image of electricity: "The electric conflict acts only on the magnetic particles of matter. All nonmagnetic bodies appear penetrable by the electric conflict, while magnetic bodies, or rather their magnetic particles, resist the passage of this conflict. Hence they can be moved by the impetus of the contending powers."

Compare those with excerpts from a paper proposing a new kind of "dark matter," by physicists Daniel Chung, Edward Kolb, and Antonio Riotto: "The goal of this paper is to show that the Universe might be made of superheavy WIMPs (we will refer to them as X particles), with mass larger than the weak scale by several (perhaps many) orders of magnitude.... To see the effects of vacuum choice and the scale factor differentiability on the large X mass behavior of the X density produced, we start by canonically quantizing an action of the form (in the coordinate $ds^2 = dt^2 - a^2(t)dx^2$)..."

The subjects of science are not only often unseeable; they are also untouchable, unmeasurable, and sometimes even unimaginable. The only way to examine these elusive entities is to scale them up, or shrink them down, or give them a familiar, solid form so that we might finally get at least a temporary handle on them. But even in 1882, physicist and lawyer Johann B. Stallo recognized that the current models of the universe were only "logical fictions," useful tools for understanding but in the end only "symbolic representations" of the real world.

When it comes to science—like so many other things—we find ourselves literally at a loss for words. Thus are metaphors born. When botanist Robert Brown first noticed the quick random motion of plant spores floating in water (now known as Brownian motion), he described it as a kind of "tarantella," according to physicist George Gamow, who went on to anthropomorphize it as "jittery behavior." (Brownian motion was the first convincing evidence for the existence of molecules, since it was bombardment by water molecules that made the plant spores dance.)

Later, Gamow described X rays as a mixture of many different wavelengths of invisible light. "Being suddenly stopped in their tracks [by a target], the electrons spit out their kinetic energy in the form of very short electromagnetic waves, similar to sound waves resulting from the impact of bullets against an armor plate." Thus in German they are called *Bremsstrahlung*, or "brake radiation."

Sometimes the metaphors get confused. A mixture of many colors is called white, but we also call a mixture of sounds "white" noise; we speak of "loud" colors. Something that is "going to seed" is deteriorating, yet "seedy" really means "fertile," since seeds are the origin of new growth. The universe is described alternately as a bubble, a void, or a firecracker. Time is "fluid," or "grainy," or both. Electrons are waves, and light waves are particles. If it all sounds as if the scientists don't know what they're talking about, it is at least in part because a lot gets lost in translation.

Imagining the unseeable is hard, because *imagining* means having an image in your mind. And how can you have a mental image of something you have never seen? Like perception itself, the models of science are embedded inextricably in the current worldview we call culture. Imagine (if you can) what the planetary model of the atom would have looked like, its satellite electrons orbiting its sunlike nucleus, if people had still thought the earth was flat. It would have been literally unthinkable. "A model or picture will only be intelligible to us if it is made of ideas which are already in our minds," wrote physicist Sir James Jeans. It was geneticist J. B. S. Haldane who noted that the inner workings of nature are "not only queerer than we suppose, but queerer than we can suppose."

Unable to suppose what the universe is really like, we rely on our rather limited but comfortably familiar models. The look of those models changes periodically, with the result that our view of the universe changes drastically. It's a long way from Newton's mechanical universe, controlled by invisible pulleys and springs, to today's image of forces as wrinkles in space, of matter as mere vibrating wisps of energy, of the physical world we know as but a shadow of a higher eleven-dimensional reality. "Scientific theories," writes Isaac Asimov, "tend to fit the intellectual fashions of our times."

Asimov goes on to detail the specific case of the atom, as good an example as any, since atoms are still essentially unseeable—or at least require a completely different kind of seeing than the one we are used to. The Greeks, who specialized in geometry, saw atoms as differing primarily in shape. Fire atoms were jagged, so fire hurt. Water atoms were smooth, so water flowed. Earth atoms were cubical, so earth was solid. Along came 1800, and the world had gone metric—in the sense of being mainly interested in measuring. Shape was no longer interesting; only *amounts* mattered. Thus atoms became featureless little billiard balls, differing mainly in the quantity of mass they contained. Later still, in the 1890s, the fashion in science was the notion of the force field—and so atoms were seen to differ mainly according to the configuration of their outer electron clouds. All these images persist today in one form or another, with physicists still focusing on quantities, organic chemists on the shapes of molecules, and so on.

Another familiar example of this phenomenon is plainly visible in the night sky. The stars in the Northern Hemisphere are clustered into constellations that mirror the images that danced in the heads of the Greeks who named them: All romance and adventure, the stars tell stories of queens and warriors, gods and beasts. The stars of the Southern Hemisphere, on the other hand, were named by a more modern culture, whose main interest was navigation. They did not see bears and lovers in the sky but rather triangles, clocks, and telescopes. "The division of the stars into constellations tells us very little about the stars," wrote Jeans, "but a great deal about the minds of the earliest civilizations and of the mediaeval astronomers."

Of course, it's not surprising that the way we see atoms and stars should change, since images of more everyday things also change drastically from time to time. Any culture's perception of childhood, the role of women, work, religion, government, all look very different in different eras. The ever adorable Judy Garland in *The Wizard of Oz* looks positively fat compared to today's child models.

Metaphors are drawn from common experiences. There is no way to imagine the unknown except in terms of the known, and so the landscape of the unfamiliar gets filled in mostly with familiar images. The images we use to describe both the unseeable subjects of science and the unseen future necessarily are fashioned from the "seeable" world we experience every day. And there's the rub. We do not experience the very large or the very small, the invisible forces and mathematical fields, the curvature of space or the dilation of time. We cannot crawl inside an atom or zoom along at the speed of light. "The whole of science is nothing more than a refinement of everyday thinking," wrote Einstein. But everyday "common sense," he also pointed out, is merely that layer of prejudices that our early training has left in our minds.

Common sense is both necessary and useful. It becomes dangerous only "if it insists that what is familiar must reappear in what is unfamiliar," writes J. Robert Oppenheimer. "It is wrong only if it leads us to expect that every country that we visit is like the last country we saw." Yet this is precisely what people do. The truth is that a model, like a foreign language, isn't really useful until you can take it somewhat for granted. It's hard to speak a language fluently when you have to keep rummaging around in the back of your mind for the right word or phrase. And it's hard to understand complicated ideas when the simple ideas and assumptions that lead up to them are still tenuous and elusive. You can't learn much about atoms if you keep having to remind yourself, "Let's see. Now, the nucleus is the thing in the middle. The electron is the much smaller thing on the outside. Is the electron the negatively charged one? Right, I remember." And so on. Being fluent means having words and ideas on the tip of your tongue. But once you become fluent in a language or in a set of ideas, you have internalized them to the extent that other languages and ideas sound automatically strange and foreign.

"Familiarity is soporific," writes physicist B. K. Ridley. It breeds consent to whatever models we're used to. It's a tender, powerful trap. "Consider the danger of familiarity," he goes on. "It seems clear that an object cannot be in two places at once; but an electron suffering diffraction can. It also seems clear that though size and position are infinitely variable, everything shares the same time; but, as Einstein showed, this is not so. We must check our intuitive ideas all the time."

It's not so easy to check these intuitive ideas, because, well, they're intuitive! Embarking on new territory requires a fresh supply of words and images. But where are they to spring from? Often unknowingly, we keep returning to the same old well. Or as Einstein put it: "We have forgotten what features in the world of experience caused us to frame [prescientific] concepts, and we have great difficulty in representing the world of experience to ourselves without the spectacles of the old, established conceptual interpretation. There is the further difficulty that our language is compelled to work with words which are inseparably connected with those primitive concepts."

In a word, language can easily turn "into a dangerous source of error and deception," Einstein said. Science has a special language problem, however, in that it borrows words from everyday life and uses them in contexts that exist only in realms far removed from everyday life. When I first tried to explain the newly discovered force particles in terms of "the force you feel when you stub your toe," I found that I had stumbled upon a semantic thicket, because "force" on a macroscopic scale and "force" on a submicroscopic scale can masquerade as very different things. Physicists borrowed the idea of force from Newton's mechanics and applied it to quantum mechanics, where it was modified—at least, to a layperson—almost beyond recognition. How can force have meaning in a system that barely allows for the notion of cause and effect? But still physicists talk about "force particles," and we who were left back with our billiard-ball images of particles and "pushing and pulling" notions of forces stay hopelessly, irretrievably confused.

"Often the very fact that the words of science are the same as those of our common life and tongue can be more misleading than enlightening," says J. Robert Oppenheimer, "more frustrating to understanding than recognizably technical jargon. For the words of science—relativity, if you will, or atom, or mutation, or action—have a wholly altered meaning."

Many physicists are particularly uneasy about terms applied to subatomic particles: "Quark" for example, was borrowed from a phrase in *Finnegans Wake;* in German it means something like "cream cheese." But "quark," to most people, doesn't mean much of anything. Far worse, say the physicists, are those words that do. The subatomic world is teeming with strange species of particles bearing oddly familiar names. "Strange" is one of them. Yet particles called "strange" or "charmed" or variously "colored" or "flavored" are not in any way particularly unusual or pleasant or green or good-tasting. The words are worse than nonsense (say some physicists), because they are downright deceiving.

Physicist Richard Feynman, for example, objected that this was "lousy" terminology: "One quark is no more strange than another quark. Maybe charm is OK, because it's so far out you know it isn't really charmed. But people think that up quarks are really turned up somehow, so it's very misleading." Victor Weisskopf concurs: "I always get the creeps when people talk about virtual particles," he says. "There is no such thing. It's a mathematical concept to describe the strength of a field." The term "virtual" refers to the very short-lived nature of such particles, but even the term "particle," Weisskopf points out, "is only there to remind you that the field has quantum effects."

It's hardly fair to pick exclusively on modern words like "charm" and "color." Where does a term like electric "charge" come from? Is it like a charge account? A charge in battle? (Obviously the usage "to get a charge out of" something comes from the science and not vice versa.) We speak of positive and negative electricity, when in fact there is no such thing—and if there were, the positive would be negative and vice versa. (Something with a negative charge actually has an excess of electrons, the particles of electric charge. Something with a positive charge has fewer electrons than it needs to make it neutral.) When an atom gets "excited," it does not sit on the edge of its seat (although it may dance around a bit). On the subatomic level, "force" means something closer to "interaction," and the strength of a force becomes the probability of its occurring.

The real trouble with words is that they automatically embody images, whether we recognize this or not. Take the word "wave," for example. It is almost impossible to think of a wave without conjuring up an image of something that looks like a water wave. And for many centuries nobody could figure out what light was, because of this linking of wave to its image in water. Water waves travel through water more or less the way sound waves travel through air and other substances. If light was a wave, it seemed painfully obvious that it had to move through something, too. The painful part was figuring out what that something might be.

As it turned out, no one could find this mysterious substance or even imagine its clearly impossible properties. It was called the luminiferous ether, and from the late seventeenth century until the time of Einstein, people were as certain of its existence as other people had been certain that the earth was flat. Yet in order to vibrate fast enough to carry light, this ether would have to have the properties of a solid. Needless to say, this posed a few problems. "If the all-penetrating ether is solid," writes Gamow,

> how could the planets and other celestial bodies move through it without practically any resistance? And, even if one would assume that the world ether is very light, easily crushable solid material, like Styrofoam, the motion of celestial bodies would bore so many channels in it that it should soon lose its property of carrying light waves over long distances! This headache was pestering physicists for many generations until it was finally removed by Albert Einstein, who threw the ether out the window of the physics classrooms.

Einstein was able to throw out the ether because he threw out the image of a light wave undulating like a water wave. A light wave could travel through nothing at all, because it is made, essentially, of a moving electric field that sets up a moving magnetic field that in turn sets up a moving electric field and so on and so forth—pulling itself up by its bootstraps. It's like an electric motor turning on a generator that turns on a motor and so on. It doesn't need to travel *through* anything because its electric and magnetic fields create each other as they zip along—at

186,000 miles per second, mind you. But it's easy to see how the image of water waves hung people up.

There are, of course, many other examples throughout history. Pythagoras's model of planets revolving on invisible spheres became so strongly entrenched in Greek thought that "the Greeks soon seemed unable to imagine any planet without its orbital sphere upon which it moved in a perfect circle," writes author Guy Murchie, "any other orbit being obviously less godly." Harvard biologist Stephen Jay Gould reminds us how hard it was for people to accept the idea of continental drift because it seemed so contrary to current thinking; once it caught on, everybody seemed to think that anyone who didn't accept it was stupid.

Einstein himself got stuck on his image of an essentially unchanging universe. He even invented something somewhat like the notorious "world ether" to make his model work. It was a mathematical device called the cosmological constant, which would oppose the pull of gravity, keeping the universe still. Later he called it "the greatest blunder of my life." Ironically, recent evidence that the universe may be accelerating at its outer edges suggests that such a repulsive force might actually exist. If so, Einstein was wrong about being wrong. Not surprisingly, physicists struggling to understand this force are inventing a new set of terms to describe it. One of these terms—quintessence—harks back to the original ether, the fifth essence (after fire, earth, air, and water).

Reductio ad Abstractium

Models are as impossible as they are perfect—just like fashion models, superwomen, or Superman. And physics is just as full of perfect but impossible things: ideal gases, perfect crystals, the ubiquitous billiard ball that serves as a model for everything from atoms to stars. A central feature of science is "the process of abstraction," writes Philip Morrison, "the distilling from some bit of the real world a more cleanly defined system that will, one hopes, still exhibit the properties of the real system in which he is interested. Much of the excitement that can be found in the practice of physical science has to do with seeking clever abstractions for complicated physical systems and then justifying the choice of the abstraction."

The abstractions of science are stereotypes, as two-dimensional and as potentially misleading as everyday stereotypes. And yet they are as necessary to the progress of understanding as filtering is to the process of perception. Science would be impossible without them—if only because the real world of nature is much too complicated to deal with in its natural form. Abstractions are a way to distill the essence from an otherwise unfathomable situation.

"Physics is about the simple things in the universe," notes one physicist, and yet "it could be argued that simple things plainly do not exist." Biology and chemistry are incredibly complex compared with physics, but even such a seemingly simple thing as a stone, he says, is "much too complicated for a physicist to deal with."

The simpler the models, the more removed they are from reality. Yet the simplest models are often the most useful ones. That's one reason that math is such a powerful tool in physics. It's the ultimate abstraction, which cleanly takes care of many of the messy details of reality by temporarily dispensing with them altogether. All models, in a sense, are intermediate steps on the road to mathematical abstraction. "The imagery allows us to move forward more rapidly, but the truth is in the math," explains Caltech physicist Kip Thorne.

As British psychologist Richard L. Gregory puts it, the images are a kind of "cartoon-language." He notes, "Just as the pictographs of ancient languages become ideograms for expressing complex ideas—finally expressed by purely abstract symbols as pictures become inadequate—so such models become restrictive. They give way to mathematical theories which cannot be represented by pictures or models."

Today mathematics has become very much the language of science. The objects of study are mathematical and so are the models and even the metaphors. I was surprised to hear theoretical physicist David Politzer of Caltech describe the most recent inventions in the physics of the early universe—the moments just after the Big Bang—as mathematical theorems. "English is just what we use to fill in between the equations," he said. "The language we use to talk to each other doesn't have analogies in nature. But we have greatly extended our mathematical vocabulary, and we are always looking to expand this set of metaphors. That's what it's all about: Understanding is a way of picturing things, and mathematics gives you a way to do it."

Politzer, like so many others, insists that the real stuff of physics is essentially nontranslatable into everyday language. But this isn't so unnerving once you consider that it's impossible to experience almost anything beyond a superficial level until you learn its special language, whether it's tennis or ballet or law. Like any other jargon, math is a vehicle that lets you go a great deal farther than you could go without it. (My friend the physicist once gave as an example of the usefulness of jargon the phrase "second cousin twice removed." Although it does not mean much unless you know the jargon of family relations, it is certainly a lot simpler to say, "Frieda is Mike's second cousin twice removed" than it is to say, "Mike is the great-great-grandson of the man who is Frieda's great-great-great-grandfather." Sailors are also well aware of the usefulness of jargon. Once I was sailing with a boatload of novices, and we were about to run aground. The skipper ordered everyone to hike out over the lee rail. About half the crew went to port while the other half scampered to starboard.)

Math is particularly useful jargon in that it allows you to describe things beautifully and accurately without even knowing what they are. You can forget about the problem of trying to imagine the unimaginable in everyday terms, because you don't need to. "The glory of mathematics is that *we do not have to say what we are talking about,*" writes Feynman (emphasis his). Curiously, these mathematical images often come closer to describing reality than images fashioned from reality itself. And as many discoveries have been made in physics by looking at equations as by looking through microscopes and telescopes. "There is a mystery to this," says Feynman, "how mathematical thinking seems to make things fit." Unfortunately (or perhaps fortunately), we cannot make a mathematics of the world, as Feynman points out, "because sooner or later we have to find out whether the axioms are valid for the objects of nature. Thus we immediately get involved with these complicated and 'dirty' objects of nature, but with approximations ever increasing in accuracy."

It is not so surprising that when the mathematical models get dressed in the metaphors drawn from everyday experience, we get into trouble. "The history of theoretical physics," wrote Sir James Jeans, "is a record of the clothing of mathematical formulae which were right, or very nearly right, with physical interpretations which were often very badly wrong." Newton's laws of motion were almost entirely right—*entirely* right if you neglect such extreme instances as travel at the speed of light. Yet when they were interpreted as the inner workings of a giant mechanical clockwork that existed in absolute space and time, they "put science on the wrong track for two centuries." In the same way, the mathematical formulas describing the interaction of electric and magnetic fields (light) went wrong only when they were interpreted as the undulation of light waves through the world ether.

Mathematical or otherwise, our images of nature are always bound to be somewhat wrong. But even inaccurate mental models can be useful. A young physicist I know believes that it is bad to introduce people to atomic structure by letting them imagine electrons in orbit around a nucleus like planets around a sun. The model is wrong, he argues. But all of us (including most scientists) begin the journey to the center of the atom with this comfortably familiar image; only later did physicists embellish it with the subtle complexities of quantum states. The orbits are a temporary framework that helped people to get their footing while climbing toward a higher level of understanding. As J. Robert Oppenheimer writes about his "house of science": "It is not so old but that one can hear the sound of the new wings being built nearby, where men walk high in the air to erect new scaffoldings, not unconscious of how far they may fall."

Scotsman James Watt constructed a workable steam engine in the eighteenth century based on an incorrect theory of heat. A hundred years later another Scot, James Clerk Maxwell, constructed a theory of electrodynamics based on "a lot of imaginary wheels and idlers in space," writes Feynman. "But when you get rid of all the idlers and things in space, the thing is OK." Physicist P. A. M. Dirac first predicted the existence of antimatter by imagining holes in empty space. Antimatter turned out to be real enough, even though the holes didn't.[3]

Sometimes models that no one takes seriously turn out to be surprisingly real. When it was first developed, the image of a force field was just a pretty picture. "In the beginning," write Albert Einstein and Leopold Infeld, "the field concept was no more than a means of facilitating the understanding of phenomena from the mechanical point of view." Yet before long the field

---

[3]Although in one sense the holes are still real, and physicists sometimes speak of "holes in the Fermi sea," and so on.

took on an unexpected reality. The physicists conclude, "The electromagnetic field is, for the modern physicist, as real as the chair on which he sits."

Models are stepping-stones. Just as Einstein built on the structure erected by Newton, so Newton built on that of Kepler and Copernicus. (If Copernicus had not published his treatise on the heliocentric solar system, if Kepler had not precisely calculated the elliptical paths of the planets, Newton could never have seen the similarity between their motions and the fall of the apple.) A model can serve as a solid foundation even when that model is assumed to be wrong. Maxwell himself apparently didn't believe in the popular model of an electrically charged atom. "It is extremely improbable that when we come to understand the true nature of electrolysis we shall retain in any form the theory of molecular charges," he wrote, "for then we shall have obtained a secure basis on which to form a true theory of electric currents and so become independent of these provisional hypotheses."

Seeing Through Images

There is a more fundamental reason for using models, however, even when these mental images of things are inevitably blurry compared to the more measured terms of mathematics. We visualize because "seeing" is inextricably linked with understanding. ("I see" is synonymous with "I understand.") Visualizing helps us think about the unthinkable. Sometimes nature is unthinkable because it's so complicated. "In one chunk of ordinary material you have 10^{23} atoms," says physicist Marvin Goldberger. "Even if you had a computer that could deal with that many interactions, you still couldn't imagine it. Even if it were practical, it wouldn't be useful. So we go back and forth, between the words and the pictures."

At other times, nature is unthinkable because it is so far removed from our everyday experience. Trying to picture the universe before the beginning of time or beyond the boundaries of space confronts us with the unimaginable. (We may be able to think about *when* the Big Bang was, for example, but *where* was it?)

Feynman understood the importance of visualization as much as anyone, despite his repeated insistence that some things can be understood only through the language of mathematics. Feynman, after all, was a master visualizer. His Feynman diagrams are a visual language for describing complex subatomic events as a collection of simpler ones. In the early 1980s, I got the rare opportunity to sit down and talk with him about how he *thought* about physics. His response was, as usual, enlightening:

> We learned something from Einstein. He wanted to put those two pieces together (electromagnetism and gravity, as part of the effort to find a common link among all the forces of nature) and he failed, partly because he started too early. It was like trying to put together a car when you only have two pieces. Today we have many more pieces of the puzzle, and the puzzle is much more complicated.

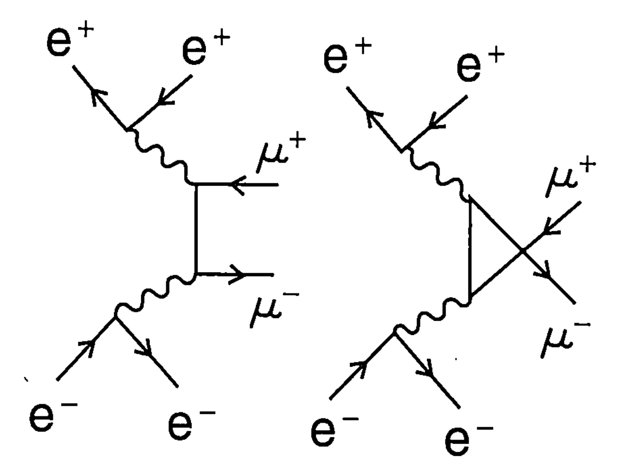

Feynman diagram.

> But he also failed—of course, I don't know why Einstein failed—but when he did his early work, he visualized a lot. A guy going up in a spaceship sending light back and forth (which helped him "see" the true relationship between time, distance, and the speed of light); a guy going up in an elevator (which helped him see that gravity and acceleration were equivalent). He got the idea that way, and then he formed elegant equations to explain the idea. He was good at it. It was nifty.
>
> Later, when he was searching for a unified theory, he used a different kind of thinking—a guessing at mathematical forms. For years I've tried that and have fallen on my face. So I'm trying to do it visually. [The Feynman diagrams] aren't sufficient for this so I'm searching for a new descriptive imagery. I'm trying to follow my own advice.

The problem that Feynman was working on at the time was the nature of the mysterious force that binds quarks and their "force particles," gluons, inside atomic particles.[4] He explained how he liked to take "the most odd, peculiar, striking" thing about the problem and abstract it away from the rest. In the case of quarks, the most striking thing is that the color force increases as the quarks get farther apart. "So you ask," said Feynman,

> What's the simplest way of stating it? How can you make a simpler problem with the same peculiarity? Say I use only two gluons, forget about the quarks. Assume that space is two-dimensional. If I try the theory with only two colors and no quarks and two dimensions, I think I understand why the gluons don't come apart. Now I have to climb back up and put the quarks back in and see if it still works. I may have thrown out the baby with the bathwater in trying to simplify, but I don't think I did.

[4]See "Forces and Pseudoforces."

Throwing out the baby with the bathwater is always a problem with building models, because models are always abstractions. One never knows for sure whether the model has really gotten to the essence of things or just gotten rid of it. And if the kinds of stereotypes we have constructed of people are any guide, the answer is not encouraging. In most cases, we have skimmed the surface while leaving the deeper reality untouched.

"To what extent do models help?" asked Feynman. "It is interesting that very often models do help, and most physics teachers try to teach how to use models and to get a good physical feel for how things are going to work. But it always turns out that the greatest discoveries abstract away from the model and the model never does any good."

Scaffolding is only a façade. Eventually even the strongest scaffolding gets cast aside and the best models are replaced by newer ones. Einstein's relativity replaced the luminiferous ether; the Bohr atom refined Ernest Rutherford's miniature planetary model by combining it with a musical image of standing waves. Scaffolding is a great support from which to build and remodel and fine-tune, but the trick is remembering that it's not the real thing.

And taking models too literally can lead us into hopeless and unnecessary confusion. People often get frustrated in their attempts to learn about atoms, because the image of everyday "particles" is so indelibly (and perhaps unconsciously) imprinted in their brains. It's natural to want to know just where an electron is—or, in the case of radioactivity, just where the electron was hiding in the nucleus before it was emitted. Just where are the electrons in an atom during the transition between one quantum state and another? (Just where are they at any time?) And which electron occupies which quantum "orbit"?

But this way of thinking about electrons, Jeans points out, is akin to believing that your bank balance actually consists of so many coins in a particular pile. When your balance changes by a certain number of dollars, you don't imagine those dollars actually flying through the air from, say, your account to that of the department stores whose bills you just paid. You do not worry about which particular dollars pay for the rent or the groceries. If you insisted on trying to put your finger on these pieces of information, you would hear yourself talking very much like a physicist talking about electrons: You would have to say that which particular dollars pay the rent is largely a matter of chance. This, says Jeans, "may be a foolish answer—but no more foolish than the question."

Models can also be misleading when used in inappropriate contexts—say, when simple models of physics are applied to complex things like people. As Stephen Jay Gould points out, "a machine makes a poor model for a living organism. Physical models often imply simple, inert objects like billiard balls that respond automatically to the impress of physical forces. But an organism cannot be pushed around so easily." Yet we talk about "the force of habit" or "the pressure to achieve," as if we were just as inert as billiard balls. We speak of "balance of power," as if we knew how much power weighed and how to measure it. We speak of "forcing" people or nations to do things, as if we knew which buttons to push to make them go and make them stop, as if there were only a single possible response to our actions.

In the end, models and metaphors are useful only to the extent that we understand their meaning and limitations. It does not do any good to understand atoms in terms of billiard balls if we don't understand billiard balls, or motives in terms of forces if we don't understand forces. As Jeans puts it, "Nothing is gained by saying that the loom of nature works like our muscles if we cannot explain how our muscles work."

CHAPTER TWO

Right and Wrong

> We may need to rely again on the influence of science to preserve a sane world. It is not the certainty of scientific knowledge that fits it for this role, but its *uncertainty.*
>
> —STEVEN WEINBERG (emphasis his), *Dreams of a Final Theory*

SOME YEARS AGO, I was invited to speak to a group of "gifted" junior high school students in our community on the subject of science and creativity. Thinking that nothing could be quite as creative as Einstein's theory of relativity (what could be more creative than refashioning our fundamental notions of matter, space, and time?), I decided to try them out on that. All went well until the end, when a girl sitting way in the back asked, "But what if Einstein was wrong?"

What indeed? It was a fair question, to be sure. Science seems littered with the mostly forgotten remnants of "wrong" ideas. Heat is not a fluid; the earth is not flat nor does it reside at the center of the universe; the planets do not revolve in perfect circles on fixed celestial spheres; Mars is not covered with canals; no luminiferous ether pervades our space, undulating invisibly as a carrier of light. On the other hand, empty space is now described as curved into four (or more) dimensions, and even vacuums are said to come in several exotic varieties. It seems that the outrageous ideas of yesterday are the scientific facts of today—and vice versa.

In the past few years alone, science has suffered a surfeit of well-publicized wrong turns, all of them fairly typical examples of science's slow zigzag toward the truth:

- A group of astronomers said that the discovery of ice on the moon was not ice at all but probably a misreading of a spurious signal—much ado about nothing.
- Some "new" planets discovered orbiting other stars faded away, as researchers challenged the data.
- The sighting of house-size snowballs from space made an enormous splash, then melted into oblivion.
- A new particle called the "leptoquark" was discovered at a particle accelerator in Europe, but now, it seems, the particle was probably a piece of stray noise.
- Two physicists announced in a prestigious physics journal that empty space spiraled like a corkscrew around some previously unknown axis, and that due to this mysterious turn of things the universe had an upside and a downside. The discovery didn't hold water, and once again scientists were left looking like people who didn't know which end was up.
- And then, there was that famous asteroid heading straight for earth—only it wasn't.

This kind of stumbling is not only inevitable but necessary. "When people are trying to do very difficult things, it's expected that some results will fall by the wayside," says Caltech astrophysicist Roger Blandford. "If people were very conservative—if they always published only what they expected to find—there would be few discoveries." Being wrong is not the worst thing that can happen to a scientist. Being "not even wrong," as the physicist Wolfgang Pauli put it, is far more devastating, because it means your idea isn't even worth disputing.

So why shouldn't Einstein be wrong?

Einstein will almost certainly be proved wrong in the long

run. Or at least wrong in the sense that he himself proved Newton wrong. But "wrong" is obviously the wrong word for it. The girl's question reminded me of a conversation I once had with MIT's Philip Morrison about whether some current views of the universe were "right" or "wrong." He said to me, "When I say that the theory is not right, I don't mean that it's wrong. I mean something between right and wrong."

Unfortunately, the territory between right and wrong is uncomfortably unfamiliar to most of us, especially when it comes to science. "It's a scientific fact" is virtually synonymous with "It's absolutely true." Smearing social theories with shades of gray is one thing, but everyone knows that scientific knowledge is black and white—or so goes the popular misconception. It turns out, however, that very little in science is actually wrong and nothing in science is ever completely right. Take Isaac Newton, for example. There is no argument about the fact that Einstein proved Newton wrong. Newton said that time and space were absolute, and Einstein proved they were not. Newton never conceived of gravity as an unseen curvature of space. Newton didn't realize that mass was a form of energy, or that inertia would become infinite as you approached the speed of light.

Yet Newton's "wrong" ideas still chart the paths of space shuttles and place artificial satellites into nearly perfect orbits. Apples still fall and the moon still orbits according to Newton's formulas. For that matter, Newton's theories work well for everything in our daily experience. They break down only at extreme velocities (approaching the speed of light), where relativity comes into play, or at extremely small dimensions, where quantum theory takes over, or in the presence of extremely massive objects such as black holes. "Einstein's correction of Newton's formula of gravity is so small," writes philosopher and novelist Arthur Koestler, "that for the time being it only concerns the specialist."

Einstein's equations even give the same answers as Newton's equations for the things that Newton was dealing with.

Einstein proved Newton wrong only in the sense that he stood on Newton's shoulders and saw things that Newton could not see—like what happens to time and space under extraordinary (to us) conditions. Mostly, Einstein proved Newton *right,* since his theories were built on Newton's foundations. Einstein took Newton's ideas and stretched them to previously unimagined limits, brought them into a new dimension, made them broader, bolder, more sophisticated. Einstein added to Newton just as today's physicists are adding to Einstein. Einstein climbed the tower of Newton's scaffolding and saw things from a better perspective. If the scaffolding hadn't been strong, he would have fallen flat on his face.

Right and wrong turn out to be surprisingly unscientific ways of describing ideas, especially scientific ideas. Rarely do revolutionary concepts overthrow old ways of thinking in unexpected coups. Physicist Hendrik Casimir goes as far as to argue that no sound theory is ever completely refuted: "There is no 'stage of refutation,' but there is all along a process of demarcation and limitation," he writes in his book *Haphazard Reality.* "A theory, once it has reached the technical stage, is not refuted, but the limits of its domain of validity are established. Outside these limits new theories have to be created."

Or as physicist David Bohm put it, "The notion of absolute truth is shown to be in poor correspondence with the actual development of science.... Scientific truths are better regarded as relationships holding in some limited domain."

New ideas expand, generalize, refine, hone, and modify old ideas, but rarely do they throw them out the window. Some "wrong" ideas are misconceived, or wrong only in that they are awkwardly formulated. Some turn out to be not so much wrong as unnecessary or irrelevant. As with the luminiferous ether, or James Clerk Maxwell's "wheels and idlers in space," or the notion that heat is a fluid, new theories render these constructs superfluous. But the misconceptions at the root of most "wrong" ideas in the history of science are sins of omission: They were wrong because they failed to take something into account, to see some part of nature that was keeping itself invisible, to notice connections among things that seem on the surface totally unconnected. "Wrong" more nearly means "limited."

For centuries, people argued over whether the wave theory of light or the particle theory of light was correct. But light turned out to be both: part wave and part particle. Both theories were right, but restricted. A correct theory requires aspects of both.

These days cosmologists engage in heated debates over whether the universe is "closed" (shaped like a sphere) or "open" (like a horn). While a closed universe will eventually collapse under its own weight, an open universe will expand forever. In a talk recently at Caltech, British physicist Stephen Hawking suggested it could be open and closed at the same time, depending on how it's sliced in eight dimensions.

Even the idea that the earth is flat was largely the result of a limited outlook on our large, spherical planet. The earth certainly *seems* flat enough as you walk around town. But the view from home is always somewhat parochial, and the earth doesn't begin to look round until you get far enough away from it. Today most people have seen the spherical earth in its true shape and colors, in images sent back by orbiting satellites. Yet hundreds and even thousands of years ago people like Columbus and Eratosthenes were able to see much the same view with the aid only of their imaginations. Physically or intellectually, the difference between a round earth and a flat one is primarily one of perspective—a broad versus a narrow point of view. Space-time itself begins to look curved only when your measurements cover a large enough territory. And quantum mechanics and relativity are merely ways of offering larger perspectives on classical ways of viewing things.

As Einstein described it, constructing a new theory is not like tearing down an old barn to erect a new skyscraper. It is rather like climbing a hill from which you can get a better view. If you look back, you can still see your old theory—the place you started from. It has not disappeared, but it seems small and no longer as important as it used to be.

Yet to gloat (or even worry) over the finding that Newton (or Einstein) might be wrong seems somewhat silly. *Of course* they were wrong. Neither Einstein nor Newton could resolve every unanswered riddle or foresee every possible consequence of every conclusion. They did not (could not) claim to be all-seeing or all-knowing. People who do claim to possess this kind of knowledge are not in the business of science, because right and wrong in that sense are not questions of science; they are only matters of dogma.

In fact, science never proves anything completely right, because there is so much left to be learned. "Each piece, or part, of the whole of nature is always merely an *approximation* to the complete truth," writes Richard Feynman (emphasis his). "In fact, everything we know is only some kind of approximation, because *we know that we do not know all the laws* as yet. Therefore, things must be learned only to be unlearned again or, more likely, to be corrected."

Or as Sir James Jeans put it, "In real science, a hypothesis can never be proved true. If it is negatived by future observations we shall know it is wrong, but if future observations confirm it we shall never be able to say it is right, since it will always be at the mercy of still further observations."

To be sure, scientists are people, and as such enjoy an aura of "rightness" as much as anyone. But from Aristotle to Einstein, the tenets of the greatest thinkers were often held much more tentatively than popular histories have acknowledged. Newton, for example, never regarded his theory of gravity as "right." Einstein remarked on the two-hundredth anniversary of Newton's death, "I must emphasize that Newton himself was better aware of the weaknesses inherent in his intellectual edifice than the generations of learned scientists which followed him. This fact has always aroused my deep admiration."

Politicians and journalists and social scientists are not so apt to admire others for admitting their mistakes; on the contrary, the admission that even part of a policy or theory is wrong is frequently touted as proof that it was (and is) completely without merit. When it comes to metaphysics at least, fixing the blame for wrong and the credit for right easily become obsessions.

In fact, the rightness or wrongness of scientific ideas tends to become tinged with dogma precisely when those ideas enter the realm of philosophy. And no wonder: Categorizing ideas as cleanly right or wrong may not be scientifically useful, but philosophically it is immensely appealing. No one likes being left in an intellectual purgatory. And so the slow evolution of scientific theories is rewritten as a series of revolutionary coups.

"Scientific revolutions are not *made* by scientists," writes Casimir. "They are *declared* post factum, often by philosophers and historians of science.... The gradual evolution of new theories will be regarded as revolutions by those who, believing in the unrestricted validity of a physical theory, make it the backbone of a whole philosophy.... Physics may even feel flattered by this homage, but it should not be held responsible for the unavoidable disappointments."

Even in science, of course, some ideas are righter than others. But how do you tell which is which? Right ideas seem to be those that lead to further investigation, to whole new categories of questions, to an even more passionate quest for knowledge. Right tends to open our eyes, wrong tends to close them. In this sense, Newton was right, but someone like Aristotle was wrong, because (as George Gamow puts it) "his ideas concerning the motion of terrestrial objects and celestial bodies did probably more harm than service to the progress of science." Galileo, among others, spent a lifetime trying to right Aristotle's wrong ideas about the immutable heavens, the geocentric universe, and so on.

But even this interpretation can be open to question. One of the things Aristotle was most "wrong" about was his assertion that all bodies naturally stop moving if they are no longer being pushed by a force. But in a passage I was surprised to find in a physics textbook, Douglas Giancoli states:

> The difference between Galileo's and Aristotle's views of motion is not really one of right and wrong.... Aristotle might have argued that because friction is always present, at least to some degree, it is a natural part of the environment. It is therefore natural that bodies should come to rest when they are no longer being pushed.... Perhaps the real difference between Aristotle and Galileo lies in the fact that Aristotle's view was almost a final statement; one could go no further. But the view established by Galileo could be extended to explain many more phenomena.

Right ideas are seeds that flower into righter ideas, whereas wrong ideas are often sterile and do not bear fruit. Once Newton got the right idea about gravity, he explained a great deal more than falling apples, or even the orbit of the moon. He tied together the universe with one cosmic force in a way that allowed later astronomers to understand the motions and masses of all the stars and planets.

Many scientists say that these *connections* are as good a guide to rightness as anything—especially when it comes to drawing lines between science and the so-called pseudosciences, such as astrology. The idea that the positions of the planets may influence your day-to-day life simply doesn't fit in with anything else people know about gravity or other aspects of nature. Any idea that seems completely unconnected with the rest of knowledge ought to be greeted with suspicion.

In the end, the importance of being wrong is greater than one might think, because a well-thought-out wrong idea serves as a basis of comparison, a springing-off point, for right ideas. Even the ubiquitous billiard ball is invaluable as a model precisely because of its obvious wrongness: "We know from the outset that it is wrong in the strict sense that it cannot possibly be true," says B. K. Ridley, "and so an assessment of how wrong it is in the particular case can begin straightaway."

Victor Weisskopf tells a story about the impatient German tourist who asks why the Austrians bother to publish railroad schedules, when the Austrian trains are never on time. The Austrian conductor answers, "If we didn't have timetables, we wouldn't be able to tell how late we are." Many scientific models, says Weisskopf, are like Austrian timetables. (Weisskopf can tell this story because he was born in Vienna, and is still an Austrian citizen.) Some of them are partly wrong and some of them are very wrong. "But what's interesting is to see *how* and *why* they are wrong," he says. "You always need the timetables."

CHAPTER THREE

Seeing Things

We're pretty good at picking out things in what looks like noise.

—University of Florida astronomer ROBERT PEÑA,
while observing on the Keck Telescope on Mauna Kea,
Hawaii, with UCLA astronomer Andrea Ghez

And sometimes we're pretty imaginative.

—ANDREA GHEZ

MANY YEARS AGO, when I first began delving into the curious Alice in Wonderland world of particle physics—that subatomic never-never land inhabited by quarks and gluons, entities strange and charmed—I asked my friend the physicist how anyone could believe in such seemingly ephemeral objects, things that no one could ever really see. And he answered, "It all depends on what you mean by seeing."

Like many people, I always feel somewhat skeptical when I hear physicists confidently claiming to have "seen" particles effervescing into existence for a mere billionth of a second, or massive quasars teetering 10 billion light-years away at the very brink of space and time.[5] I know for a fact that they have seen no

[5]Quasar is short for quasi-stellar object. Quasars are sources of immense outpourings of energy found far away from us in space and time. Their exact nature is still unknown.

such thing. Quarks and quasars are invisible to the naked eye. At best, the physicist has seen a bump on a curve plotting the ratio of various kinds of particles produced in a subatomic collision or the faint fingerprints left by 10 billion-year-old photons on silicon detectors; more often, such a "sighting" is in truth a conclusion laboriously drawn from long hours of computer calculation and long chains of inferences and assumptions. Hardly the sort of thing to inspire an exultant "Eureka!" (Or even "Land ho!") Sometimes the things that scientists say they see are so removed from actual quarks or quasars that one wonders if they (or we) should believe their eyes.

Physicists "see" exotic particles by bombarding them with other particles and analyzing the patterns created as the particles bounce back into their electronic detectors. The first person to "see" the atomic nucleus used much the same method, except that the electronic detector used was the human eye. During World War I, Ernest Rutherford aimed a beam of particles streaming from a radioactive rock toward a thin sheet of gold foil. Most of the particles passed right through. But some—surprisingly—were scattered through very large angles and a few were even reflected *backward.* "It was about as credible," said Rutherford, "as if you had fired a fifteen-inch shell at a piece of tissue paper and it came back and hit you." From this Rutherford concluded that the atom was not a uniform mass, as previously thought, but something rather more like a miniature solar system, with almost all of the mass concentrated in a small, central nuclear "sun." Most of the particles passed right through because there was very little inside the gold atoms for them to hit; if a particle did chance to hit the nucleus, however, it could be reflected backward like a ball hitting a brick wall.

Today, physicists similarly "see" all kinds of particles by bombarding all kinds of targets with all kinds of other particles,

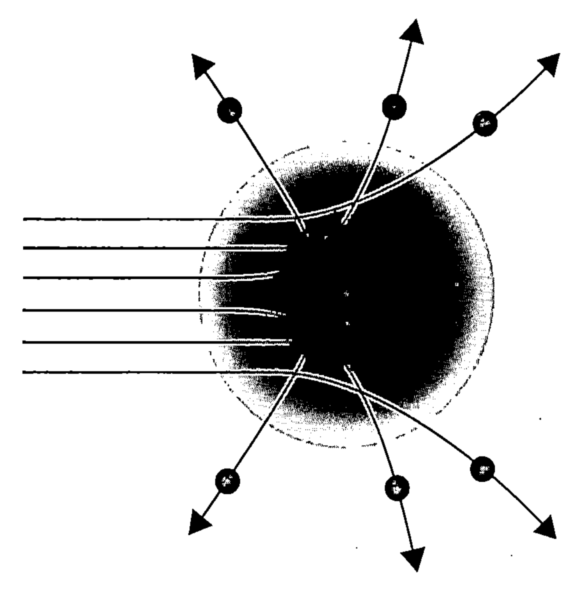

How Rutherford "saw" the nucleus.

and they use sophisticated electronic equipment to analyze their results. But in truth, seeing inside an atom is not so different from seeing a friend or a building. It only *seems* unreal and abstract, because the details of the process are slightly less familiar: The way we go about seeing things in everyday life is scarcely what you'd call "direct."

I see this page, for example, because rapidly vibrating atoms in the filament of the lightbulb overhead and in the sun outside my window some 93 million miles away are sending out streams of light particles called photons, some of which are showering down right now on the paper's black-and-white surface. A few of these photons collide (as in an atom smasher) with the molecules in the ink, and are absorbed; others hit the pigment in the paper and are redirected back in the direction of my eyes. If they penetrate the pupil, they will be focused by a lens onto a light-sensitive screen (the retina), a sophisticated electronic detector that passes along information about the photons' energies, trajectories, and frequencies to my brain in the form of digital bits. On the basis of laborious calculations and long chains of inferences and assumptions, my brain concludes that the light patterns represent printed words, conveying some rough translation of the writer's passing thoughts.

Of course, my eyes, like particle detectors and telescopes, are tuned in only to the narrowest band of information coming from the outside world. The pupil is but a tiny porthole in a sea of radiation. In a universe alight with images, we are mostly in the dark. Human eyes respond only to those electromagnetic vibrations that make waves between .00007 and .00004 of a centimeter long. Yet, as I type, I am bombarded by other kinds of electromagnetic waves as small as atoms and as large as mountains, coming from the far reaches of space, from the inside of my own body, from the radio transmitter twenty miles away. I know that these signals are there, in the room with me, because if I flip on the radio or television I will suddenly be able to see or hear them, in the same way that visions suddenly "appear" before me the minute I open my eyes. If I had still other kinds of detectors (I can sense some of the infrared radiation on the surface of my skin as heat), I could pick up still other kinds of signals. Yet we walk through this dense web of radiant information every day without being the least aware of its existence.

Radiation is only one kind of information to which we are mostly blind. We are deaf to most of the sound around us. Our chemical senses (taste and smell) are extremely limited compared to those of a plant, or a cell, or a dog. We can barely perceive the difference between hot and cold: a blindfolded person can't tell whether she has been burned by a hot iron or dry ice. Even our perception of forces is curtailed by our size. While we can easily sense the pull of gravity, we are almost completely insensitive to the pulls and pushes of air resistance and surface tension that are

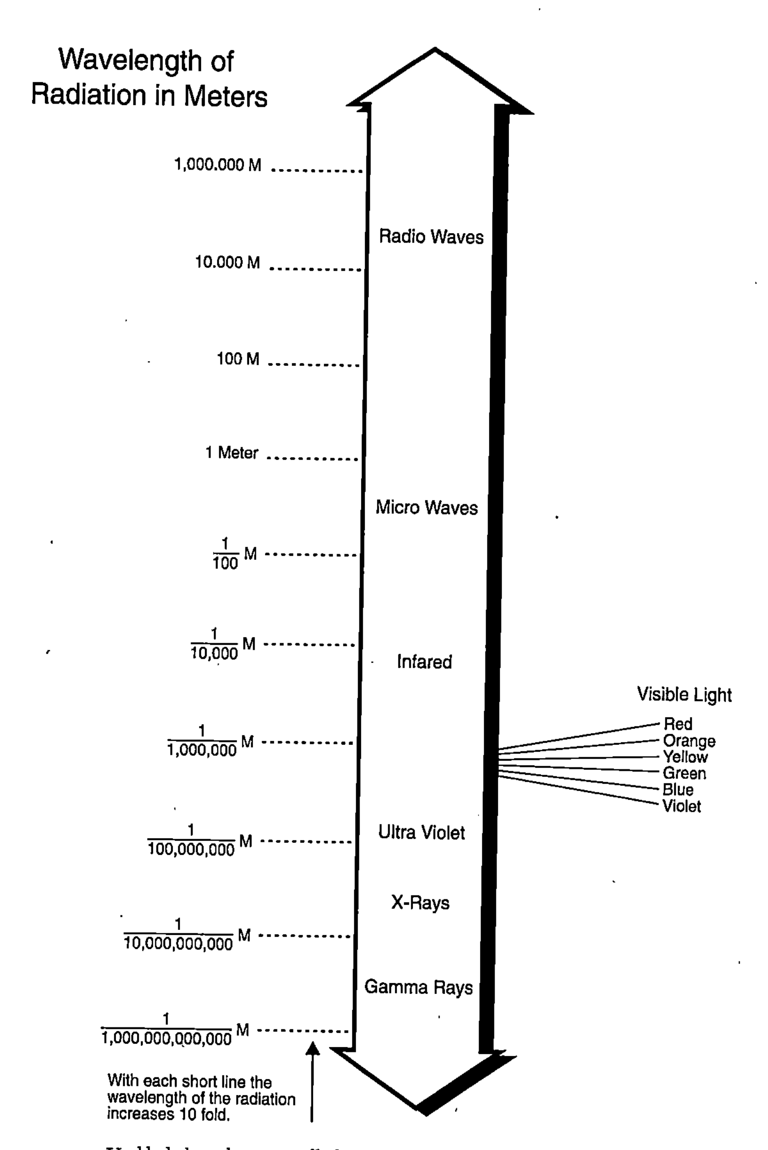

Visible light is but a small sliver of the electromagnetic spectrum.

major forces in the lives of cells and flies. We don't have to push the air out of the way to walk through it, as the gnat does; on the other hand, the electrical force of cohesion is relatively so much stronger for a small creature that the fly can crawl up the wall, completely ignoring gravity. To say that we are narrow-minded (or at least "narrow-sensed") is the least of it.

Naturalist Loren Eiseley writes of coming upon a spiderweb in a forest. The spider is confined to its own two-dimensional universe, totally oblivious to the plants or people around it, even to the pencil Eiseley uses to reach out and pluck it: "Spider thoughts in a spider universe—sensitive to raindrop and moth flutter, nothing beyond, nothing allowed for the unexpected, the inserted pencil from the outside world."

We live in a spiderweb, too—a three-dimensional spiderweb spun in the unseen context of our four-dimensional space—and we are only beginning to become aware of the vast universe outside. Our perceptions of time and space are largely limited to things in our own experience, of our own relative size. We find it almost impossibly difficult to comprehend numbers much larger than those we can count on our own fingers and toes, or spans of time much longer than our lifetimes. Looking outside our spiderweb takes an enormous leap of the imagination.

From our spider point of view, the world is clearly flat. It is also obviously motionless. It was probably Galileo who first proposed the idea that we cannot tell by experiment alone whether we are moving or not. Put yourself inside the closed cabin of a steadily moving ship, he said. Allow small winged creatures like gnats to fly around, watch fish swimming in a bowl, toss objects back and forth, notice how things fall to the ground. No matter how many experiments you perform, "you shall not be able to discern the least alteration in all the forenamed effects, nor can you gather by any of them whether the ship moves or stands still."

What you perceive as "standing still" at the equator is in reality a rapid spin around the earth's axis at the dizzying rate of 1,000 miles per hour. In addition, the entire spinning earth is whizzing around the sun with a speed of almost 20 miles per second. The solar system itself is moving with respect to the center of our galaxy at 120 miles per second, and our galaxy, the Milky Way, is rushing toward the neighboring Andromeda Galaxy (from its point of view) at 50 miles per second. And that's not all: If you looked at the earth from a far-off quasar, you might see us speeding away at 165,000 miles per second, close to the speed of light.

What Galileo stumbled upon in stepping out of his spiderweb was what Einstein would later refine into the theory of relativity. Einstein saw that there were other things we couldn't perceive, like the elastic nature of space and time. We were being deceived by our senses into thinking that our own Euclidean three-dimensional geometry was the geometry of the universe. Einstein was out of his senses. He saw that it didn't matter whether or not our own crude perceptual instruments could pick up the tiny increases in the mass of objects as they moved faster. It was a limitation of our instruments. Today's particle accelerators routinely push particles to speeds almost as fast as the speed of light, and see them gain forty thousand times their initial weight in the process!

In addition to all the things that we *can't* see are the things that we *don't* see because we choose to ignore them. Right now, I am choosing to ignore the sounds of my own breathing, the feel of the ring on my finger, the sight of the glasses that are right in front of my nose—and even the nose on my face. Shutters and pupils are meant not to let information in but to shut it out. As anyone knows who has ever held a camera, too much information is easily as blinding as too little. If you played all nine Beethoven symphonies at once, you would hear nothing but noise.

But deciding what to turn on and off and when is a dangerous business—especially since we are mostly unaware of it. Recently, four people sat in my living room directly underneath a loud antique clock. At 3:05 I asked whether the clock had struck three. Two people insisted it had and two insisted it hadn't.

Our eyes automatically erase the oversize feet and out-of-focus images that are clear to the more objective camera's eye. Distractions erase reams of information, which is one reason there is a difference between "listening" and "hearing." It is a physical fact that you cannot listen to even two conversations at once, or see in sharp focus more than the narrowest sliver of a visual field.

Sorting information from "noise" is one of the most important processes in all of perception. Yet it is also obviously a minefield of potential mistakes. There is a simple and striking illusion in which two facing profiles suddenly appear as a vase, which just as suddenly can fade again into two facing profiles. You cannot see both vase and faces at once, because you cannot see something as background and foreground (or information and "noise") at once. Whatever is in the background becomes as invisible as if it weren't there, even though you may be looking right at it.

Mostly what disappears are the sights we get used to—like our noses and eyeglasses, but also our sunsets and even the sounds of our children. Steady signals fatigue our senses, numbing our powers of response. Dogs will sleep through all kinds of everyday noises only to snap into alertness at the soft step of an intruder; parents have been known to snore through sirens and garbage trucks only to awake alert at the merest whimper from their newborn baby.

Sensory atrophy is largely learned. But some is automatic. That is, some kinds of signals actually fatigue our physical sensors to the point that we are no longer able to perceive them. Perhaps the most common example of this is the afterimage. If you stare at a bright light or open your eyes to the streaks of sunlight coming through the edges of the shades first thing in the morning, you are likely to look away only to see the image lingering in your field of vision. The afterimage is dark where the original image was bright, because it corresponds to those places on your retina where the sensors have been bleached by the light. For some moments, they can no longer respond. They can no longer send the signal to your brain saying, "white wall here" or "blue sky here." The rest of your retina responds normally, so what you see is a normal background with a dark image of the original bright "flash" imposed on it.

Some painters actually color their work with afterimages in mind. For when your eyes tire of one color, they will see its complement. Say, for example, you stare for fifteen seconds or so at a bright-red area on a painting or on a wall. The red sensors in your eye fatigue. If you then look at a white wall, your eyes will send the following message to your brain: white *minus* red. Since white minus red is green, green is the color you see. (If you stare at a green spot, you see a red spot when you look at the white wall. And so on.)

Your motion sensors work much the same way. If you spin around the room in one direction, your motion sensors soon lose their ability to respond to the steady signals: They no longer send a message to the brain that you are turning clockwise; turning clockwise has become synonymous with "stopped." When you stop spinning, the fatigued sensors respond by sending a message to the brain that says, "no longer stopped; spinning in opposite direction." You sense yourself spinning counterclockwise. If you stare at falling water for fifteen seconds or so and then switch your gaze to the ground, it can make the ground seem to "fall up," a phenomenon appropriately known as the waterfall effect.

Sensory fatigue can sometimes cause you to perceive the *opposite* of the signals you actually receive.

People who are brilliant scientists (or writers or parents or doctors or carpenters) are those who have a special talent for keeping the important things in focus—both separating the signal from the noise and also knowing when what sounds like noise might contain the quiet whisper of important information.

The instruments of science have vastly extended our senses. Indeed, physicist David Bohm concludes that "science is *mainly* a way of extending our perceptual contact with the world," its purpose being to foster "an awareness and understanding of an ever growing segment of the world with which we are in contact." Technology has unveiled vast new vistas, opening up untapped realms of time, space, and temperature. To modern telescopes and particle accelerators, the radio waves and gamma rays invisible to us are rich with images. The number of so-called elementary particles has proliferated wildly because the instrumentation to "see" them has gotten better and better. The same is true of the number of stars in the sky, and such strange newcomers to the galactic zoo as pulsars and quasars and probable planets around stars other than our sun. We can see out into space, back into time, inside our own genetic structure. We can see what the stars are made of and how a virus looks. We can measure things smaller and larger, colder and hotter, faster and slower than could ever be "seen" before. With the help of high-powered computers, scientists can extrapolate to the end of the universe or the beginning of time or the center of the earth. They can "see" what happens when chemicals react, particles collide, hurricanes evolve.

"How rich we are," writes Guy Murchie, "that we can look on these worlds with the perspective of modern science . . . that we do not have to wonder as did former men whether stars are jewels dangling from celestial drapery or peepholes in the astral skin of creation!"

Our view of the universe is changing so rapidly partly because our ability to see is growing so rapidly. That's one reason why ideas that seem right today get overturned so readily tomorrow. The more we see, the more we correct our vision. "Early descriptions of the universe are egocentric and based on the physical size and capabilities of man," writes Richard Gregory in his marvelous book on perception, *The Intelligent Eye.* Prescientific philosophy was based solely on human perception. But now we know that there is a lot going on that we can't see *except* through science. "The simple fact that stars exist invisible to the unaided eye," Gregory writes, "made it unlikely that the heavens are but a backcloth for the state of human drama."

Scientific perception has a different authority from personal perception, because it can more easily be shared. It's a way of seeing that many people can agree on—or at least agree on a way of thinking about. But the process is essentially the same: Scientists "see" by gathering data, measuring, making assumptions, and drawing conclusions. "Elementary particles don't seem real to ordinary people, because they aren't perceived in an ordinary way," says MIT physicist Vera Kistiakowsky. "Something like astronomy *seems* more real, because you can see the stars with your own eyes. But even that is mostly inferred. All science involves the interpretation of secondary information."

All *perception* involves the interpretation of secondary information. We are always seeing a great deal more than meets the eye. The light patterns that form on the tiny screens within our eyes are upside down, full of holes and splotches, badly bent out of shape. Most of what we see is in our heads. If I believed my eyes, I would see people shrink to Thumbelina size as they walked away. In fact, all my visions would remain inside my body. For it is our brains that perform the incredible feat of projecting what we see "out there" fróm the backs of our eyes to some arbitrary place in space. Not only vision but *all* sensory experience takes place within our bodies. Yet we attribute these properties to objects that exist outside us. We say that ice cream tastes sweet, or the table feels hard, when in fact it is *we* who taste and feel.

Galileo recognized that qualities such as color and smell "can no more be ascribed to the external objects than can the tickling or the pain caused sometimes by touching such objects." We are tickled or hurt not by the feather or pin but by the interpretation of an electrical signal within our brains.

"We each live our mental life in a prison-house from which there is no escape," writes Sir James Jeans. "It is our body; and its only communication with the outer world is through our sense organs—eyes, ears, etc. These form windows through which we can look out onto the outer world and acquire knowledge of it."

Our newfound scientific senses are even farther from direct interpretation. The images of quasars seen by radio telescopes using Very Long Baseline Interferometry, for example, are really composite patterns resolved from information recorded separately at individual antennas as much as six thousand miles apart, synchronized by atomic clocks and pieced together by computers. They are not "images" in the ordinary sense but rather interference patterns, like the rapidly moving moirés you see when two fences or fireplace grates or curtains overlap—secondary patterns emerging from the combination of two (unseen) patterns.

Human science no longer experiences the world through human senses. Indeed, much of scientific knowledge these days completely contradicts our senses, which is why it is so difficult to accept such concepts as quantum mechanics and curved space. The sights and sounds and objects and motions around us are not divided up into small quantum bits, like still frames from a moving picture. The space around us does not seem to bend or change or curve. "This has led to a curious situation," writes Gregory. "The physicist in a sense cannot trust his own thought." And yet, he points out, we have to learn with the "nonperceptual concepts" of physics: "We are left with a question: How far are human brains capable of functioning with concepts detached from sensory experience?"

The answer has to be that there is more than one valid way of seeing things. If we listen to Bach with our ears, and then "listen" again with (other kinds of) electronic detectors, we will pick up very different sets of signals. Both kinds of perception are equally indirect. There are many possible windows on reality. Indeed, one of Einstein's most radical notions had to do with the multiple realities of space and time; that is, the space or time we perceive depends on the means we use to mark it off, on our point of view. "Space has no objective reality except as an order or arrangement of the objects we perceive in it," writes Lincoln Barnett, "and time has no independent existence apart from the order of events by which we measure it."

Perception, after all, is a very *active* process. We do not just sit around waiting for information to rain down on us. We go out and get it. In the process, we alter it and even create it. One of the strangest things about the way physicists "see" elementary particles is that they often create them out of the energy of other particles to make them visible—something that doesn't seem quite "fair." But as Philip Morrison points out, you can't see the rapidly rotating blades of a fan unless you stop them or throw a rock at them. You can't sense radio waves by putting your hand in front of them, but you can if you tune your receiver so that it vibrates in resonance with the incoming signal. The sound waves coming from the radio are as much created in the process of detecting them as the particles created in accelerators.

What we see depends on what we look for. It also depends

on our point of view. A house viewed from an airplane does not look at all the same as a house viewed from its own front door, or from the window of a rapidly passing car. A baby does not recognize a toy viewed from the top as the same toy that looked so very different when seen from the side. A rotating shadow of any three-dimensional object will take on an amazing variety of different shapes. Which is the "true" perspective? It may be that the only wrong perspective is the one that insists on a single perspective. Like the baby with the toy, we may be mistaking one thing for two. Space and time, energy and matter, waves and particles, are all different aspects of the same thing.

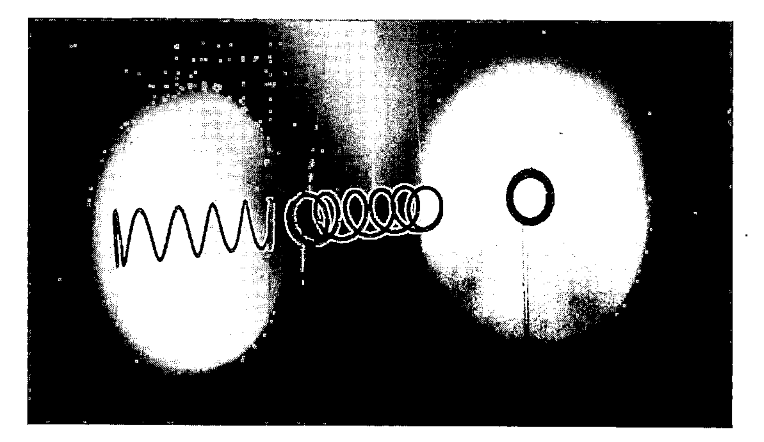

One shadow of the spring is a circle; another shadow of the same spring is a wave. If you could perceive the spring only by looking at its shadows, you might easily come to the conclusion that it was two different things.

Gathering, arranging, and sorting the information from the outside world is only the first step, however. We still must decide, What is it? What does the information mean? Meson or proton? Streetlight or moon? Shadow or burglar? Planet or star? A small dim object close up looks the same as a large bright object from afar. Stick out your thumb and use it to size up a house far away outside the window. The house may seem no bigger than the tip of your nail. Then how do you know it isn't? "When a tiny meteor smaller than a pea is falling through the air," writes Jeans, "it will send the same electric currents to our brains as will a giant star millions of times larger than the sun and millions of times more distant. Primitive man jumped to the conclusion that the tiny meteor was really a star, and we still describe it as a shooting star."

You have no way of knowing that there is anything behind you until you turn around, and yet we are not surprised to see things there. You have no way of knowing that your next step will fall on solid ground, and yet you take it on faith. Sometimes we are fooled: The piece of fluff floating a few feet away looks like an airplane miles up in the sky (or vice versa). Yet most of the time we are remarkably accurate in the way we size up our familiar world and get around in it.

"Familiar," however, is the key. As Richard Gregory points out, perception is a matter of seeing the present with images stored from the past; it is a matter of selecting the most likely (that is, the most familiar) object, the most "obvious" answer for the question, What is it? "This acceptance by the brain of the most probable answer implies a danger: It must be difficult, perhaps somewhat impossible, to see very unusual objects," Gregory notes. And if perception is a matter of making sense of the world with our limited collection of answers from the past, he asks, "then what happens when we are confronted with something unique?"

The answer is, we don't see it. The seventeenth-century Dutch scientist Christiaan Huygens drew detailed pictures of the planet Saturn as seen through his homemade telescopes. But he never recognized the unusual patterns as the now familiar rings. He couldn't see the rings largely because he wasn't *expecting* to see rings. "We not only believe what we see," writes Gregory, "to some extent we see what we believe.... The implications about our beliefs are frightening."

Gregory is hardly the only scientist to come to this conclusion. Loren Eiseley writes, "Each man deciphers from the ancient alphabets of nature only those secrets that his own deeps possess the power to endow with meaning." Sir Peter Medawar, in *Advice to a Young Scientist,* dismisses the notion that scientific discoveries are made by "just looking around":

> I myself believe it to be a fallacy that any discoveries are made in this way. I think that Pasteur and Fontenelle would have agreed that the mind must already be on the right wavelength, another way of saying that all such discoveries begin as covert hypotheses—that is, as imaginative preconceptions or expectations about the nature of the world and never merely by passive assimilation of the evidence of the senses.... The truth is *not* [emphasis his] in nature, waiting to declare itself.... Every discovery, every enlargement of the understanding begins as an imaginative preconception of what the truth might be.

In fact, it is often hard to disentangle the image of the "real" world from the preconceptions we project on it. How much about the universe do we actually discover and how much do we impose? Often it is not clear which we are doing—like the person who heard a lecture on astronomy and afterward complained that while the lecturer had beautifully explained how the scientists had discovered the sizes and temperatures of the stars, he had neglected to tell how the scientists discovered their *names.*

Perceptions are at best good guesses. They are the shortcuts we need to keep the information coming in from the outside world from becoming overwhelming, but they are never (necessarily) the truth. We have to judge a book by its cover, because we don't have the time or the resources to look inside every book. We assume that if we see a face looking out of a window there is a body attached to it, and that something that looks like a tree or a sailboat or a star probably is. We recognize patterns that seem familiar in a wide variety of contexts; even a child seems to know instinctively that a Saint Bernard is a "woof-woof" along with a Pekingese. But it is important to remember that it is we who have decided that these two very dissimilar creatures are both "dogs," in the same way as we call both the tool in the kitchen and a configuration of stars a "dipper." It is we who see the "man" in the moon. Snap judgments are as useful and essential as they are treacherous and misleading.

"We cannot... believe in raw data of perception and suppose that perceptually given 'facts' are solid bricks for basing all knowledge," writes Gregory. "All perception is theory-laden."

We see what is familiar and we see what we choose to see, which are often the same thing. If people engage in conversation only to disagree later completely on what was said, then in fact they really "heard" very different things, according to current memory research. "Memory is not a literal recording," says Harvard psychologist Daniel Schacter. "It's more like a kind of [evolving] sculpture." Laboratory studies have revealed consistent biases in the ways we remember things. For example, memory often acts as an ego booster, according to University of Washington memory expert Elizabeth Loftus: "We remember we gave more to charity than we [actually] did, we took more airplane trips, we had kids who walked and talked earlier than they did, we voted in more elections."

What we learn to see is also culturally conditioned, and this applies even to the things we see through the supposedly objective eyes of science. When people first look through a microscope, they often have trouble seeing anything but their own eyelashes, or random specks of light. They have to be taught how to see an amoeba—just as they "have to be taught" racial prejudice.

Culture also affects what scientists "see." Its influence is so strong that when a brilliant supernova appeared in 1054, at the height of religious belief in Europe, not a single European chronicle mentioned it, though it shone for many months. It was not recorded because it was not considered important. It was filtered out of history as effectively as the bonging of my clock was filtered out of my living room.

Galileo first got into trouble when he had the temerity to see such a supernova. "The appearance of a new star in heaven," writes George Gamow, "which was supposed to be absolutely unchangeable according to Aristotle's philosophy and the teachings of the Church, made Galileo many enemies among his scientific colleagues and among the high clergy." Galileo's telescope revealed (among other things) that Venus and Mercury, like the moon, sometimes have crescent shapes, implying that they orbit the sun. But what he saw was "certainly more than the Holy Inquisition could permit; he was arrested and subjected to a long period of solitary confinement."

We need not go back in history to see examples of this cultural perceptual conditioning. Who would think that civilized people are so culturally accustomed to rooms with right angles that they would rather see a person shrink before their eyes than a room change shape? Yet this happens in the famous Distorted Room invented by the psychologist Adelbert Ames. The room is misshapen in such a way that a person viewing it through a thin slot (which makes binocular depth perception impossible) has to make a perceptual choice: Either people inside shrink and grow as they move about or else the room isn't the usual rectangular shape. People always choose the right angles. Even our perceptual belief that things in the distance don't actually shrink is to some extent culturally conditioned. A psychologist visiting The Exploratorium in San Francisco told the staff about a forest-dwelling tribe of pygmies who had no experience seeing things at long distances. One day this tribe spied a herd of cattle, or some such animal, on a far-off plain—and of course assumed that they were tiny creatures, perhaps the size of ants. Imagine the pygmies' terror when the tiny creatures grew to great size as they approached!

The Impossible Triangle appears impossible only because we insist on seeing it in the familiar shape we call a triangle. What is truly impossible is the ability to perceive totally unfamiliar objects.

It's fun to be amused by optical illusions. But illusions can be disillusioning. Usually the only illusion involved is a departure from our usual point of view, a contradiction of our preconceived notion of reality. Perceptual expectations not only make it impossible to see the shape of things correctly, they can lead us to see "impossible" things. A case in point is the Impossible Triangle, a structure of three thick beams joined at right angles. Of course, no triangle can consist of three right angles, but the Impossible Triangle is no ordinary two-dimensional triangle. It is a configuration of three beams joined at right angles that *looks* like a triangle from a particular point of view. It is only impossible as long as we insist on seeing it as a familiar triangle, instead of the unfamiliar figure that it really is. The unfamiliar figure, on the other hand, appears impossible, "*although it exists,*" says Gregory (with emphasis!).

We dismiss the real as impossible and experience the impossible even though it cannot be real. If you soak one hand in ice water and one hand in hot water, and then plunge both hands into a bowl of warm water, one hand will feel cold and the other hot, even though they are both resting in water of the same temperature. On the other hand, if you touch metal and Styrofoam at the same temperature, the metal will feel much colder. What you perceive is "impossible" only because your understanding of temperature is incomplete.

Intellectual knowledge alone does not change what we perceive. The Impossible Triangle, the Distorted Room, the experiment with the hot and cold water retain their power to fool us even after we have learned how they work. We still see the sun "rise" and "set." We see the stars hang from a flat ceiling, earth stretched out flat. We see the moon as a disk about a foot across and a mile away, even when we know it is a sphere about 240,000 miles away and 2,000 miles in diameter. Look at the sun sometime and try to see it as a star 93 million miles away. It is truly impossible. "It is an effort to get perceptual and intellectual knowledge to coincide," writes Gregory. "If the eighteenth-century empiricists had known this, philosophy might have taken a very different course. No doubt there are also implications for political theory and judgment."

The moon in particular is a "heaven-sent object for perceptual study," says Gregory. The size and distance error in our perception of it are about a millionfold. But what is truly surprising, he says, is that we attribute any size or distance to it at all. We have no frame of reference. We have no experience with objects so large or so far away. Then how do we know the brightness of stars, the size of quasars, the distance of far-off galaxies? The answer is, we make long chains of assumptions. If even one of those assumptions is just a little bit wrong, our perceptual conclusion will be a great deal wrong.

In fact, the moon, as almost everyone knows, changes size in the sky as our assumptions about its distance from us change. The moon looks larger on the horizon than it does when it's high in the sky. There is no widely accepted explanation for this, but one theory is that the "ceiling" of the sky seems closer to us than the far-off horizon. Therefore the same-size object, if it were closer, would have to be smaller to make the same-size image in our eyes. A basketball across the room and a Ping-Pong ball at arm's length make the same-size image on your retina. If you didn't know the size of the basketball, you would automatically see it as larger than the Ping-Pong ball, merely because it was farther away. When we assume the moon is farther away on the horizon, we see it as larger, too.

In addition, we all have a wealth of experience that tells us that things flying by us—planets, balls, planes, birds—all get smaller as they recede into the distance. If they *didn't* seem to get smaller, then it could only be because they were actually growing bigger! The same thing happens with the moon. It doesn't really grow bigger, but it does seem to recede in the distance. And since its image doesn't get smaller, you can only assume that it, too, is growing larger.

If this seems hard to believe, you can prove it to yourself with a simple experiment. The next time a bright light flashes in your eyes, look at your hand and you may see a small afterimage of the object floating on your palm. If you then look at a wall a few feet away, and then at a wall even farther away, the afterimage will get bigger and bigger. It helps to blink to keep the afterimage in view; your brain tries to erase the "extraneous" image by pushing it aside, but blinking can bring it back. The size of the actual image is imprinted on your retina and stays the same. But your brain automatically makes it appear larger or smaller, depending on the presumed distance of the "object" making the image.

You can also alter the size of the moon merely by changing your perspective. When you see the "large" moon on the horizon, if you look at it upside down—say, through your legs—so that the perspective of the horizon disappears, it will suddenly appear "small" again. Often we have to turn things upside down to see them in a proper perspective.

There are still other limits to our perceptual powers: Some of them have to do with the fact that we are a part of the nature we study (how clearly can you think about thinking, for example?); others have to do with the fact that if you look at something closely enough, you always have to disturb it. Behavioral scientists are constantly plagued by this problem, but so are physicists. There is no way to simultaneously "see" the exact position and velocity of a subatomic particle, because in measuring one aspect, you automatically disturb the other. In trying to look at subatomic things, we are, as Lincoln Barnett says, somewhat in the position of a blind person trying to discern the shape and texture of a snowflake. As soon as it touches our fingers or tongue, it dissolves.

Some people have interpreted these perceptual limitations to mean that objective reality does not "really exist"—whatever that means. But the world is full of many important and lovely things that elude measuring. Love, for one. But also hate, humor, and almost all other emotions, ranging from our reaction to the sight of blood to our uncanny response to the arrangement of sounds we call music. If you try to dissect the emotional power of a painting bit by bit, it will dissolve (at least temporarily) as surely as a snowflake.

In any event, "reality" means different things to different people. As physicist Max Born put it, "For most people the real things are those things which are important for them. The reality of an artist or a poet is not comparable with that of a saint or prophet, nor with that of a businessman or administrator, nor with that of the natural philosopher or scientist."

This does not mean, Born says, that our sense impressions are some kind of "permanent hallucination." On the contrary, we can often agree on the nature of objective reality, despite its many appearances: "This chair here looks different with each movement of my head, each twinkle of my eye, yet I perceive it as the same chair. Science is nothing else than the endeavour to construct these invariants where they are not obvious."

Our scientific perception of reality grows and gains confidence in the same way that a baby gains confidence in the everyday reality of his or her world. We not only make hypotheses, we *test* them (or we should). If the same things happen enough times, we gain confidence in our theories. If the rattle falls to the floor every time it rolls off the table, this is unlikely to be a coincidence. If what we see with our eyes is confirmed by our sense of smell or hearing or touch, so much the better. And if other people seem to sense the same things we do, the more convincing still. We cannot eliminate the subjective aspects of perception, but we can subdue them. "It is impossible to explain to anybody what I mean by saying, 'This thing is red,'" writes Born, "or 'This thing is hot.' The most I can do is to find out whether other persons call the same things red or hot. Science aims at a closer relation between word and fact."

David Bohm concluded that what we perceive is what is *invariant*—what does not change under altered conditions. The baby learns to perceive the bottle correctly when he or she learns that it does not change when it's seen from various points of view. And it was the realization that the speed of light is invariant that led Einstein to his special theory of relativity. The surprising conclusion that space and time were relative came from the much more fundamental insight that the laws of nature (like the speed of light) were invariant under all conditions.

The strength of our beliefs is buoyed by the connections between our observations and beliefs. The more threads we can tie together, and the more tightly they are knit, the less likely it is that something major can slip unnoticed through a perceptual blind spot. Not only does the rattle fall to the floor but the moon itself is falling toward the earth and the earth toward the sun; rivers and rain and cold air all sink to the center of the planet. Gravity gains force as more and more phenomena can be explained by it. Therefore even though there are many things that astrophysicists or subatomic physicists cannot really "see," even in theory, they gain confidence in their solutions to puzzles as more and more pieces fit. If some underlying idea were fundamentally wrong, writes Victor Weisskopf, "our interpretation of the wide field of atomic phenomena would be nothing but a web of errors, and its amazing success would be based upon accidental coincidence."

Our innate perceptual limitations are as necessary as they are sometimes deceiving, so there's no point in wishing them away. We will simply have to accept, as British astronomer Sir Arthur Eddington observed, that any true law of nature is likely to seem irrational to rational human beings. Still we can do what scientists do to diminish the margin of error: We can try to get to know our instrument, its calibration, position, limitations, frame of reference—in short, how it works.

That is, we can better get to know our perceptual selves. No longer is perception the relatively simple matter it was in the Galilean world, where the universe was considered an object under inspection. Now we know we are part of that universe, and any valid observation has to take the workings of our inborn instrumentation into account. As Niels Bohr put it, science is a fascinating adventure in which we are not only spectators but actors as well.

CHAPTER FOUR

The Scientific Aesthetic

> POETS SAY science takes away from the beauty of the stars—mere globs of gas atoms. Nothing is "mere." I too can see the stars on a desert night, and feel them. But do I see less or more? The vastness of the heavens stretches my imagination—stuck on this carousel my little eye can catch one-million-year-old light. A vast pattern—of which I am a part—perhaps my stuff was belched from some forgotten star, as one is belching there. Or see them with the greater eye of Palomar, rushing all apart from some common starting point when they were perhaps all together. What is the pattern, or the meaning, or the *why*? It does not do harm to the mystery to know a little about it. For far more marvelous is the truth than any artists of the past imagined! Why do the poets of the present not speak of it? What men are poets who can speak of Jupiter if he were like a man, but if he is an immense spinning sphere of methane and ammonia must be silent?

This poetic passage appears, of all places, as a footnote in a physics textbook—volume I of *The Feynman Lectures on Physics,* to be exact. It is a most eloquent argument that science, if anything, *adds* to an aesthetic appreciation of nature. Science does not strip nature bare of its emotion and beauty, leaving it a naked set of equations. On the contrary, a scientific understanding of nature can deepen the awe, expand the sense of mystery.

Science and art may seem unlikely partners, but they have

been cohabiting as long as humans have been trying to make sense of the world around them. Robert Root-Bernstein, author of the book *Discovering*, has concluded from a host of studies that scientists and artists tend to be much more similar to one another than to other professionals, such as merchants and lawyers.

It's no surprise that artists and scientists are drawn to the same subject matter. Nature—and human nature—is rich in the kinds of mysteries that attract both. A tree is fertile ground for the botanist *and* the poet. Painters, psychologists, sculptors, and physicians all study the relationship between mother and child and the structure of the human form. Mathematicians and artists are drawn to the symmetry of snowflakes, sine waves, the spiral growth of shells. Physicists, philosophers, and composers explore the origins of the universe, the nature of life, and the meaning of death.

Yet when it comes to approach, we are told, the affinity breaks down completely. Artists approach nature with feeling; scientists rely on logic. Art elicits emotion; science makes sense. Art, like child rearing or an interest in social welfare, is supposed to require a warm (if not bleeding) heart. Science, like law or manufacturing, is supposed to be rational, objective, deductive. Scientists are supposed to think, but artists are supposed to care.

In fact, nothing could be farther from the truth. It would be a poor poet indeed who was uncontrolled, and no scientist ever got very far by sticking exclusively to "the scientific method." Though for some reason it seems to upset people, scientists are frequently every bit as passionate about their work as artists are about theirs.

Take Charles Darwin, for example. While he was rummaging around the Galápagos Islands gathering the evidence that would eventually lead to his theory of natural selection, he was hardly what you'd call detached. "I am like a gambler and love a wild experiment," he wrote. "I am horribly afraid. . . . I trust to a sort of instinct and God knows can seldom give any reason for my remarks.... All nature is perverse and will not do as I wish it. I wish I had my old barnacles to work at, and nothing new." And so on.

Such passion was also rampant in the early days of the quantum debate. Albert Einstein said that if classical notions of cause and effect had to be renounced, he would rather be a cobbler or even work in a gambling casino than be a physicist. He based his objections on what he called "an inner voice," and wrote to Born that "the theory accomplishes a lot, but it does not really bring us any closer to the secret of the 'Old One.'" Niels Bohr called Einstein's attitude "appalling" and accused him of "high treason." Another major contributor to quantum theory, Erwin Schrödinger, said, "If one has to stick to this damned quantum jumping, then I regret having ever been involved in this thing." Expressing a similar sentiment, physicist Enrico Fermi threw up his hands at the great numbers of elementary particles appearing in the 1950s: "If I could remember the names of all these particles, I would have been a botanist." On a more positive note, P. A. M. Dirac talked about his discovery of the equation that eventually led to the discovery of antimatter as "the most exciting moment of my life." And Einstein talked about the universe as a "great, eternal riddle" that "beckoned like a liberation." Hardly the stuff of your stereotypical dispassionate scientist.

As George Sarton put it in his *History of Science:* "There are blood and tears in geometry as well as in art.... It would be very foolish to claim that a good poem or a beautiful statue is more humanistic or more inspiring than a scientific discovery; it all depends upon the relation obtaining between them and you."

Yet somehow the idea that scientists are supposed to be objective about their work has come to include the idea that they are not supposed to *care* about it. "What a strange misconception has been taught to people," my friend the physicist once remarked. "They have been taught that one cannot be disciplined enough to discover the truth unless one is indifferent to it. Actually, there is no point in looking for the truth unless what it is makes a difference."

Great artists and great scientists are often known for combining *both* approaches in their work. Artists need a scientific (or at least technical) knowledge of the materials they use—paints, paper, marble, lenses, strings, computers, and so on. A musician may know as much as a physicist about resonance, acoustics, and harmony; a photographer, as much about the properties of light. Even the content of the artist's work is more often than not based on the scientist's perception of nature; after all, it was not so long ago that science was "natural philosophy." Art, like all forms of perception, is conceived in a cultural context, and much of that context is construed from the perspectives people have on their physical world.

Scientists, for their part, rely on an artistic approach for those all-important leaps of the imagination called insight. Without it, they would be forever stuck in their past perceptions. Science alone logically leads us smack into our inability to imagine (much less recognize) the unknown. "This is the power of art," writes Richard Gregory: Pictures can "make us see things differently by changing our object-hypothesis." They can expand our reservoirs of possible realities.

"Each time we get into a 'log-jam,'" writes Feynman, "it is because the methods that we are using are just like the ones we have used before. The next scheme, the new discovery, is going to be made in a completely different way. So history does not help us much." Deduction takes you only to the next step in a straight line of thought, which in science is often a dead end. As Feynman concludes, "A new idea is extremely difficult to think of. It takes fantastic imagination."

The direction of the next great leap of imagination is guided as often as not by the scientist's vision of beauty. Einstein's highest praise for a theory was not that it was good but that it was beautiful. His strongest criticism was "Oh, how ugly!" He often spoke and wrote about the *aesthetic* appeal of ideas. "Pure logic could never lead us to anything but tautologies," wrote the French mathematician Henri Poincaré. "It could create nothing new; not from it alone can any science issue."

Poincaré describes the role that aesthetics plays in science as that of "a delicate sieve," an arbiter between the telling and the misleading, the signals and the distractions. Science is not a book of lists. The facts need to be woven into theories, like tapestries out of so many tenuous threads. Who knows when (and how) the right connections have been made? Sometimes the most useful standard is aesthetic. Schrödinger refrained from publishing the first version of his celebrated wave equations because they did not fit the then-known facts. "I think there is a moral to this story," Dirac commented, in a now-famous quote. "Namely, that it is more important to have beauty in one's equations than to have them fit experiments.... It seems that if one is working from the point of view of getting beauty in one's equations, and if one has a really sound insight, one is on a sure line of progress."

It is the aesthetic sense that prompts scientists and other people as well to insist that an idea "looks" right, or "feels" wrong. Physicist Andrew Strominger, who leans far out on the limb where eleven-dimensional space meets black holes, says he's often guided by a sense of what "smells" right. He works by trying to sniff out the most aesthetic answer to a question: "[Another physicist] would say, If A and B, then C. And he'd set out to prove it. I would say, Gosh, wouldn't C be nice? And then set out to prove it."

As Weinberg put it, "the physicist's sense of beauty is also supposed to serve a purpose—it is supposed to help the physicist select ideas that help us to explain nature."

Sometimes, the connection between art and science can be even more direct. Niels Bohr was known for his fascination with cubism—especially "that an object could be several things, could change, could be seen as a face, a limb, a fruit bowl," as a friend of his later explained. Bohr went on to develop his philosophy of complementarity, which showed how an electron could change, could be seen as a wave, a particle. Like cubism, complementarity allowed contradictory views to coexist in the same natural frame.

Some people wonder how art and science ever got so far separated in the first place. The image of Einstein with his violin is almost as familiar as the image of Leonardo with his inventions. It is a standing joke in some circles that all it takes to make a string quartet is four mathematicians sitting in the same room. Even Feynman played the bongo drums. (Although he found it curious that while he was almost always identified as the physicist who played the bongo drums, the few times that he was asked to play the drums "the introducer never seems to find it necessary to mention that I also do theoretical physics"—something he attributed to the fact that people may respect the arts more than sciences.) Still, Feynman was also known to remark that the only quality art and theoretical physics have in common is the joyful anticipation that artists and physicists alike feel when they contemplate a blank piece of paper.

In truth, the definitions of both art and science have narrowed considerably since the days when science was natural philosophy. According to social theorist Sir Geoffrey Vickers in Judith Wechsler's excellent collection of essays *On Aesthetics in Science,* before this unnatural separation everyone knew that knowing was an art, that understanding required skill, that both art and science were process as well as product. Science was restricted to its present narrow meaning only at the end of the nineteenth century, when the term came to apply to "a method of testing hypotheses," based on experiments. Conducting experiments to test theories has little to do with art. But Vickers suspects that the difference is deeper. People want to believe that science is a rational process, that it is describable. Intuition is not describable and should therefore be relegated to a place outside the realm of science. "Because our culture has somehow generated the unsupported and improbable belief that everything real must be fully describable," writes Vickers, "it is unwilling to acknowledge the existence of intuition."

There are, of course, substantial differences between art and science. Science is written in the universal language of mathematics; it is, far more than art, a shared perception of the world. Scientific insights can be tested by the good old scientific method. And scientists have to try to be dispassionate about the conduct of their work—at least to the extent that their passions do not disrupt the outcome of experiments. Sometimes, inevitably, they do disrupt them. "Great thinkers are never passive before the facts," says Gould. "Hence, great thinkers also make great errors."

The history of science seems to be written in these errors—primarily because the discoverers of the workings of the phyical world have been propelled as much by emotion as by thought, by conviction as by deduction. Johannes Kepler eventually proved that the orbits of the planets were elliptical and not round as everyone had thought—thereby laying the ground for Newton and Einstein—but Kepler never abandoned his belief in the notion that the orbits of the six known planets were enclosed in the five perfect geometrical solids. The way he thought about astronomy is hardly what you'd call logical; in fact, he went as far as to argue that the orbits that lie outside the earth's orbit "have it in their nature to stand upright," whereas those inside the

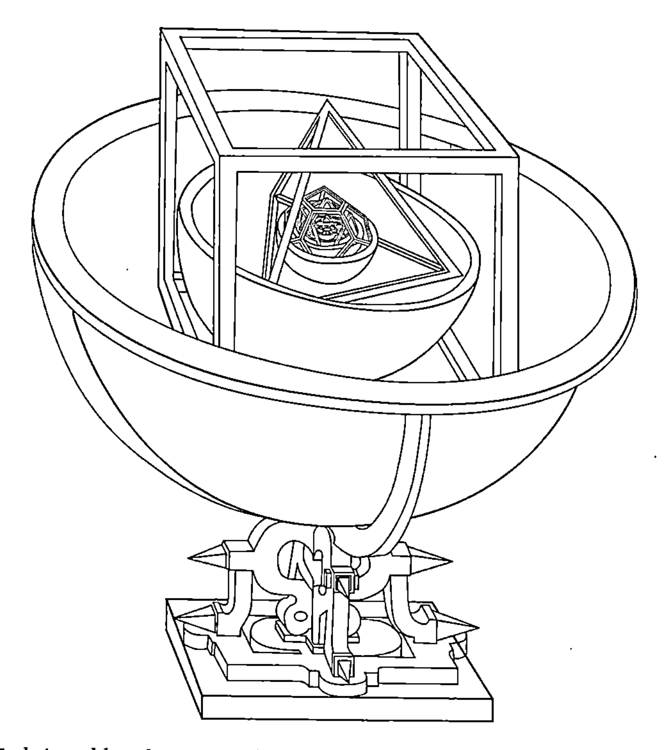

Kepler's model of the universe, showing how he thought the orbits of the planets were positioned with relation to geometric shapes. Redrawn from Mysterium Cosmographicum *(1596, edition of 1621).*

earth's orbit have it in their nature to float. "For," he explained, "if the latter are made to stand on one of their sides, the former on one of their corners, then in both cases the eye shies from the ugliness of such a sight."

This leaves the question: If scientists are as often as not guided by a certain aesthetic, by their notion of beauty, what does "beauty" mean to a scientist? Obviously, it does not mean "pretty" or "pleasing" or even "inspiring" in the normal sense. An equation or a theory could not be considered a thing of beauty as long as we stuck to this rather narrow definition. It turns out that beauty in science is something much closer to the notion of *simplicity.*

"You can recognize truth by its beauty and simplicity," writes Feynman.

> When you get it right, it is obvious that it is right—at least if you have any experience—because usually what happens is that more comes out than goes in.... The inexperienced, the crackpots, and people like that, make guesses that are simple, but you can immediately see that they are wrong, so that does not count. Others, the inexperienced students, make guesses that are very complicated, and it sort of looks as if it is all right, but I know it is not true, because the truth always turns out to be simpler than you thought.

Simplicity is beautiful to a scientist because it is a surprisingly pervasive aspect of nature and has a great power to lead to important ideas. How many things that seemed so wildly disparate in the past have turned out to have common threads, intimate connections? The forces of nature, electricity and magnetism, falling apples and orbiting planets. As poet Muriel Rukeyser observed, even islands are connected underneath.

"It is simple, and therefore it is beautiful," says Feynman of gravity.

> It is simple in its pattern. I do not mean it is simple in its action—the motions of the various planets and the perturbations of one on the other can be quite complicated to work out, and to follow how all those stars in a globular cluster move is quite beyond our ability. It is complicated in its actions, but the basic pattern or system beneath the whole thing is simple. That is common to all our laws; they all turn out to be simple things.

The appeal of higher dimensions is that it makes things simpler, whether those are the four dimensions of space-time or the eleven-dimensional string theories. A cartoon in a physics journal depicted a professor writing an enormously long and complex set of equations on the blackboard, to the horrified stares of baffled students. "Don't worry," the professor says, trying to calm their fears. "It looks a lot simpler in ninety-six dimensions." The simplicity comes from the clarity of understanding, from the ability to see through the distractions and focus on the essential elements, to explain many seemingly unconnected things with one "simple" idea. According to Weinberg, beauty in physics is "the sense of inevitability... the sense that nothing in the work could be changed."

Nature seems to be built on patterns, and looking for those patterns is the primary preoccupation of artists and scientists alike. "What's beautiful in science is that same thing that's beautiful in Beethoven," says Weisskopf. "There's a fog of events and suddenly you see a connection. It expresses a complex of human concerns that goes deeply to you, that connects things that were always in you that were never put together before."

In the end, the connections between science and art may rest on the matter of motivation. When MIT metallurgist Cyril Stanley Smith became interested in the history of his field, he was surprised to find that the earliest knowledge about metals and their properties was provided by objects in art museums. "Slowly I came to see that this was not a coincidence but a consequence of the very nature of discovery," he writes, "for discovery derives from aesthetically motivated curiosity and is rarely a result of practical purposefulness."

CHAPTER FIVE

Natural Complements

> We cannot at the same time experience the artistic content of a Beethoven sonata and also worry about the neurophysiological processes in our brains. But we can shift from one to the other.
>
> —VICTOR WEISSKOPF

WEISSKOPF LIKES to tell the story about a conversation that took place many years ago between Nobel Prize winners Felix Bloch and Werner Heisenberg. The two famous physicists were walking along the beach, and Bloch was sounding out Heisenberg—so the story goes—on the significance of some new theories having to do with the mathematical structure of space. At length Heisenberg responded, "Space is blue and birds fly in it."

The story is especially nice because it illustrates so well what many physicists believe to be the most profound contribution of quantum theory, the enormously successful edifice that explains the atom and everything it composes and comprises. This contribution was not a discovery in the normal sense: not a particle, not a new kind of extraterrestrial object or event, not even a theory or equation. It was, rather, a philosophical outlook that allowed scientists to see beyond the mass of paradoxes that seemed to be making modern physics all but impenetrable. It was the notion of complementarity. And if nothing else, the story of Heisenberg and Bloch captures the essence of complementarity:

that one can talk about the same subject in two very different kinds of terms, and that what makes good sense in one context can make absolutely no sense in another.

Complementary ideas are opposing ideas that add up to much more than the sum of their parts. They complement each other like night and day, male and female. Complements are required for a full spectrum of understanding, just as a full array of colors is required to produce pure white. (In fact, any two colors are considered complementary if they add up to white.) Complements are the yin and yang of science. Or as the physicist Emilio Segrè wrote, "It is one of the special beauties of science that points of view which seem diametrically opposed turn out later, in a broader perspective, to be both right."

Complementarity is almost everyone's favorite when it comes to "the sentimental fruits of science." And no wonder: Life, like nature, is fairly bursting with unresolved—unresolvable—paradoxes. Particles are waves and waves are particles. You are ninety-eight cents' worth of cosmic star dust floating at the obscure edge of an ordinary galaxy, and yet you are the center of your own world; to friends and family, you may be precious beyond all worth. On one day, humanity seems the apex of all things beautiful, generous, mindful; another day, it seems a stupid beast. But the truth is that we are both beauty *and* the beast—just as energy is only another way to look at matter.

Danish physicist Niels Bohr fathered complementarity as a way to tame the inherent limits to precise measurement on the scale of atoms. For example, in the quantum realm, one cannot pinpoint the position of a particle without sacrificing knowledge of its motion, or measure its motion without introducing uncertainty about its position. Bohr said that the reality of particles required *complementary* descriptions, more than one point of view. It doesn't matter that you can't measure both motion and position at the same time; you can't see both sides of a coin at the same time, either. As long as people insist on viewing the subatomic world with their everyday perspectives (and what choice have they?), they will be stuck with looking at nature one dimension at a time. "In our description of nature," said Bohr, "the purpose is not to disclose the real essence of phenomena but only to track down, as far as it is possible, relations between the manifold aspects of experience."

The idea of complementarity reaches far beyond uncertainties in measurement. It also helps to explain the wave/particle duality of light. By the turn of the century, decades of argument and experimentation had finally convinced physicists that light had to be a wave. It bent, or diffracted, around corners as an ocean wave bends around a reef. Two sets of light beams produced the bright and dark interference bands that result when waves reinforce and then cancel each other. Then Einstein came along and showed that light came in parcel-like clumps. Light was "quantized." Things seemed so contradictory that one physicist was moved to remark that nature behaved according to quantum theory on Mondays, Wednesdays, and Fridays, and wave theory on Tuesdays, Thursdays, and Saturdays.

It seemed obvious to everyone that light could not be *both* waves and particles. The notion of a wave is as different from the notion of a particle "as is the motion of waves on a lake from that of a school of fish swimming in the same direction," says Weisskopf. A particle is like a bullet—a patently material, finite object that occupies a particular place in space and time. A wave is more like a motion—a continuous abstract form. If the wave theory was right, then the particle theory had to be wrong. If one was truth, the other was by definition heresy.

It turns out, of course, that both are right. Waves and particles are complementary ways of describing the nature of light, just as position and motion are complementary aspects of particles. Not only light but also all energy and matter and radiation reveal the same curious duality. Electrons can be diffracted like light beams, and show the same interference patterns when passed through the layers of aligned molecules in a crystal.

The thing that's hard to swallow about complementarity is that the various aspects of an entity needed to make it "whole" can also be mutually exclusive—like waves and particles. Complements are more than a fancy physicist's version of "on the one hand/on the other hand." When you look at one side of a coin, you may not be able to see the other side of the coin, but the unseen side does not seem impossible or absurd or cease to exist. Waves and particles, though, seem to present incompatible choices. They do not make sense in the same context, and it is hard to see how both can describe the same thing.

The trick to complementarity is knowing when which view is appropriate. For as J. Robert Oppenheimer pointed out, "The more nearly appropriate the first way of thinking is to a situation, the more wholly inappropriate is the second."

The statement that "space is blue" is hardly an appropriate way of expressing a mathematical relationship—but then an equation would hardly be an adequate way of describing one's sense of the sky during a stroll on the beach on a summer's day. The more a particle acts like a particle, the less it behaves like a wave—just as the clearer the position of a particle becomes, the more fuzzy its motion becomes (and vice versa).

In the most extreme cases, focusing sharply on one aspect of a situation actually destroys the other. This idea is embodied in the notorious Heisenberg uncertainty relations. Werner Heisenberg himself visualized the problem this way: If you want to measure the exact location of an electron and also its exact motion, you run into a problem. In order to "see" it, you have to shine some kind of light on it. If you use low-energy light, so as not to disturb the motion of the electron, the wavelength of the light is so long that it will not be able to define the electron's position. (It's analogous to using a coarse-tooth comb when you need a fine-tooth comb.) On the other hand, if you use high-energy light (a fine-tooth comb), you can determine the position of the electron accurately, but the light will give it such a jolt that it will change the particle's motion.

Einstein never liked the uncertainty principle, because he didn't like the notion that there were things we couldn't—not even in principle—measure. But what Einstein saw as an intellectual dead end, Bohr and others saw as a philosophical treasure trove. The answers were turning out to be limited only because the questions were inappropriate. "The Heisenberg relations are warning posts that say, 'Use ordinary language only up to here,'" says Weisskopf. "When you get to the dimensions of the atom, you get into trouble."

The uncertainty principle, in other words, may result mainly from a misapplication of a metaphor. In the atomic realm, notions like "location" and "motion" may not make sense. As Sir James Jeans put it, "It is probably as meaningless to discuss how much room an electron takes up as it is to discuss how much room a fear, an anxiety, or an uncertainty takes up."

So the uncertainty is not as spooky as it seems. It means only that you cannot chemically analyze your piece of cake and eat it, too. Because by the time you chemically break down the cake to analyze it, the cake will have dissolved into something else entirely. "Every intervention to make a measurement, to study what is going on in the atomic world," writes J. Robert Oppenheimer, "creates... a new, a unique, not fully predictable situation."

A quantum state, like a ballerina's arabesque or a bird's song, retains its character only as long as it remains whole. This is not to say that you cannot break down these things into their individual notes and motions and molecules, only that watching an animal in the wild and dissecting it in a laboratory are complementary ways of exploring nature.

The inherent duality means that any view of nature or human nature that views one "side" as dogma and the other as heresy is probably wrong, or at least dangerous. As Weisskopf points out, the domination of one idea has inevitably led to abuses—whether it was the dominance of religious dogma during the Middle Ages or the excessive influence of technology today. "Whenever one way of thinking has been developed with force, claiming to encompass all human behavior, other ways of thinking have been neglected," says Weisskopf. "This has its roots in a strong human desire for clear-cut, universally valid principles containing answers to every question. But because human problems always have more than one aspect, general-purpose answers do not exist." The answers are not either/or but "all of the above"—or at least "each" of the above at times when they are appropriate.

MIT professor of computer science Joseph Weizenbaum made much the same point about the danger of overreliance on computers. Society's love affair with computers, Weizenbaum said, is a symptom that the scientific mode of thinking is becoming "imperialistic." Not that scientific thinking is bad, only that it becomes dangerous when it overwhelms all other approaches. He notes, "If you wanted to understand the Great Depression of the 1930s and you only looked at Department of Labor statistics and you didn't read novels by people like John Dos Passos because novels are not scientific, then that's bad—because in a very deep way you can learn more by reading the novels."

For centuries people argued over whether light was essentially a wave or essentially a particle. Today this seems as superfluous as arguing about whether space is blue or whether it has mathematical properties. Each, in its proper context, is true. This doesn't imply that the whole truth lies somewhere in between the two viewpoints: Complementarity is not a compromise; it is rather like the sides of a box or the facets of a problem. What you see depends on what side of the box you look at, which is why light—and in fact all energy and matter—shows up as quanta in some experiments and behaves like waves in others.

As such, complementarity makes it easier to accept the innate limits on perception, measurement, and the extent to which we can imagine the unseeable world. Each way of seeing goes only so far. Like the image engraved on the back of each eyeball by incoming light beams, all models and perceptions are flattened representations of reality. Like science and art, each complementary mode of seeing and interpreting the world adds vital elements that the other cannot.

A microscope enhances your ability to see, but only at the expense of the larger context. If you put a living organism under a microscope, you can see far more clearly into its individual membranes and cells. But even then, as Weizenbaum points out, "it wouldn't make sense to say that what you're seeing in any way resembles the essence of the organism itself."

Accepting complementarity merely means accepting the idea that because one view is right, the opposite view isn't necessarily wrong—that truth is not the other side of heresy (or vice versa). If people are somehow coming to the conclusion that science is a one-way street to a single right answer, that would be profoundly ironic, for as Max Born wrote: "This loosening of the rules of thinking seems to me the greatest blessing which modern science has given us.... The belief that there is only one truth and that oneself is in possession of it seems to me the deepest root of all the evil that is in the world."

PART II:

Movers and Shakers

There is no sense in regarding matter and field as two qualities quite different from each other.... What impresses our senses as matter is really a great concentration of energy into a comparatively small space.

—EINSTEIN AND INFELD, *The Evolution of Physics*

CHAPTER SIX

Forces and Pseudoforces

> If you insist upon a precise definition of force, you will never get it!
>
> —Richard Feynman, *Lectures on Physics*

Nowhere have the words of physics infused the everyday vernacular so much as in the realm of forces—the influences that propel people to do things, or that leave them perpetually stuck in ruts. We speak of people who gravitate toward certain interests, who are pressured to succeed or bogged down by inertia, who cause friction in groups or organizations. We describe people as magnetic, forceful, repulsive, or even electrifying. We talk of being pushed into doing things, of being attracted by places, people, or jobs. We even talk of "pushy" people. Most of the time we think we know what we are talking about.

Physicists, on the other hand, tend to be much more cautious when speaking in terms of forces. J. Robert Oppenheimer noted that even Newton, who probably came up with more formulas for forces than anyone else, never understood what a force really was. "Was this . . . something that spread from place to place, that was affected only instant by instant, point by point; or was it a property given as a whole, an interaction somehow ordained to exist between bodies remote from each other? Newton was never to answer this question."

In Newton's time, the great philosophical mystery about forces was how they got from there to here. How did an influence spread over distance? Especially distances in empty space? How could you make something move if you couldn't reach out and touch it? Or push it with a stick? Or pull it with a rope?

In the mid-nineteenth century, British experimentalist Michael Faraday came up with the notion that these forces might be transmitted by something akin to rubber tubes that stretched between two poles of a magnet, or two opposite electric charges; he even speculated that a similar mechanism might transport the pull of gravity. Scotsman James Clerk Maxwell formulated these ideas into the precise equations that described electromagnetic force fields. A field is a kind of extended aura that surrounds a body, spreading its power to affect things, not unlike political spheres of influence.

Today physicists are more likely to talk of forces as carried by special kinds of particles. These force particles often materialize out of the bursts of energy created when other particles collide in giant accelerators. The familiar photon, or light particle, is just such a force particle. But all this talk of "force particles" makes one wonder whether there is any difference anymore between the actual stuff of matter and the pulls and pushes that make it come and go.

In 1983, two of these force particles—the W and Z particles that carry the so-called weak force involved in radioactivity—were found at CERN (the European Organization for Nuclear Research), in Switzerland. The discoveries were hailed in the front pages of the *New York Times* and were widely discussed in popular science magazines. But to people who are familiar with the forces of wind or tide, of springs and hammers, of fire and chemistry, of muscles and jet propulsion, of friction and magnets and gravity and static electric charges, this all seems very puzzling, to say the least. Or as someone finally said to me in total frustration after reading one of these articles, "I wish someone would explain these forces to me in terms of the force I feel when I stub my toe."

Unfortunately, while people certainly know a great deal about what forces *do*, they know a great deal less about the mechanisms that make them work. Explaining the *what* is easy; untangling the *hows* and *whys* is a tricky thing.

In one sense, it doesn't matter a whit whether a force seems to act like a field or like a particle, whether it seems a discrete event or a property of space. All this is imagery on which different kinds of thinking can be based. What matters is what actually happens. Lo and behold, even in physics much of the terminology of forces has been altered since the introduction of quantum mechanics. Forces between particles are described much more accurately as "interactions." When two particles interact with each other, they exchange energy and/or momentum. This is exactly what happens when you stub your toe. The energy from the egg you ate this morning (which came from the sun by way of the chicken that ate the corn that absorbed the sun and then laid the egg) is converted into the electrical energy of nerves, which is converted into the kinetic (or motion) energy of your muscles. When your toe hits the doorsill, some of that energy is exchanged with the doorsill. The energy of your kick heats up the molecules in the doorsill, and the doorsill kicks back some of the energy in the form of pressure you feel as pain.

Forces are obviously very real and important things, but describing them in the metaphors of everyday life can leave us somewhat muddled. A force particle no more pushes other particles around than charmed particles are charming. Strictly speaking, a force is a transfer of energy and momentum. Two objects interact and things are never quite the same for either of them again. What happens in between is not readily describable. As

Bertrand Russell put it, a force is something like a sunrise—a convenient way of explaining something. A force no more literally forces something to happen than the sun literally rises. Electricity "is not a thing, like St. Paul's Cathedral; it is a way in which things behave. When we have told how things behave when they are electrified, and under what circumstances they are electrified, we have told all there is to tell."

Forces, in other words, are really ways of describing the way things are *connected* to each other. Inertia, action/reaction, relative forces, and fundamental forces all relate one part of the universe to another part of the universe. And the way we describe them, it turns out, really *does* matter.

Sir Arthur Eddington points out the difference that descriptions can make in his story of two special fish called Isaac and Albert. Isaac and Albert were flatfish, living in a sea of two dimensions. Isaac looked at the motions of the other fish around him and noticed that all of them seemed to curve from their normally straight paths in a curious way. He discovered that the "force" behind their curvature was another, larger fish, a sunfish, who attracted all the other fish to him. As Guy Murchie retells the story:

> This adequately accounted for most of the peculiar curves, so nobody bothered about the lesser attractions of a small moonfish circling nearby or the great numbers of fixed starfish twinkling in the background. And the only discontentment left was the carping of a few carp who did not see how the sunfish could exert such great influence from such a distance, though they presumed his influence must spread forth somehow through the water.

Then along came Albert, who said that the fish were not attracted to the sunfish by *forces* so much as they were forced to swim around the sunfish in curves, because the sunfish was sit-

A flatfish named Albert saw that there was an unseen mound in the structure of space.

ting (or swimming, as the case may be) on top of a large mound. Of course, the other fish could not directly sense such a mound, because a mound has three dimensions and the fish were two-dimensional. In the same way, we three-dimensional creatures are unable to sense the curvature of our own space, which includes the fourth dimension of time. And in the same way, it is the geometry of space-time—the "mounds" in it caused by the presence of mass—that results in what we have previously described as "forces" like gravity.

It turns out that you can actually calculate and measure the slight differences that would result depending on whether Isaac or Albert was right. And in several major tests, Albert's theories have been confirmed. So it does matter whether the various parts of nature are knit together through fields, forces, particles, and/or geometry. Or, as the real Isaac wrote in his *Principia* back in 1686:

> I am induced by many reasons to suspect [that all the phenomena of Nature] may... depend upon certain forces

by which the particles of bodies, by some causes hitherto unknown, are either mutually impelled towards one another, and cohere in regular figures, or are repelled and recede from one another. These forces being unknown, philosophers have hitherto attempted the search of Nature in vain; but I hope the principles here laid down will afford some light either to this or some truer method of [natural] philosophy.

Pulling on the Stars

What a remarkable idea, that when you accelerate into a run, your muscles are fighting the influence of galaxies scarcely visible even with the most powerful telescopes!

—B. K. Ridley, *Time, Space and Things*

One of the most puzzling forces to confront physics is technically not a force at all. It's resistance to force, or *inertia.* Inertia is simply a resistance to a change in motion. It's harder to toss a bowling ball than a tennis ball because the bowling ball has much more inertia. Inertia means an unwillingness to be pushed around, a proclivity to keep right on going the way you were going—or if you were stopped, to stay stopped. You can call it a habit, or a rut. But whatever you call it, it's not strictly a force. Newton defined force as an action exerted on a body in order to change its state; he said that inertia was a measure of the *resistance* to that change in state. In this sense, force and inertia seem to be opposites.[6]

Yet inertia feels exactly like a force when your car, for example, comes to a sudden stop and the "force of inertia" keeps you going through the windshield. Force and inertia may seem

[6]Of course, Newton also recognized that this resistance was itself a powerful force, in his famous law of action and reaction.

opposite in their origins, but they are often identical in their effects.

Newton referred to inertia as the "innate force of matter." The more matter there is, therefore, the more inertia it naturally harbors. You can pull the tablecloth right out from under the table settings because inertia keeps the heavier glasses and silver anchored in place on the table. It took a rather brilliant insight on the part of Galileo to recognize inertia at all. Previously Aristotle (among others) had assumed that the natural tendency of things was to come to a stop, and that planets (among other things) needed a constant push to keep them going. The difference between Aristotle's view and Galileo's view was that Aristotle thought the natural state of things was at rest. Galileo realized that this natural state could also be motion.

Aristotle's was a very natural assumption. After all, cars, people, and even runaway roller skates eventually come to a stop if someone or something doesn't supply the energy to keep them going. In fact, *everything* in the everyday world eventually settles into a "natural" state of rest. What Aristotle didn't realize was that there was a hidden force behind that tendency to stop: the force of friction. (And if you consider friction to be part of the natural state of things, then Aristotle was right after all.)

Many people don't realize that it takes as much energy to bring something to a stop as it does to get it going. Space probes visiting distant planets not only have to carry enough fuel to get them where they're going, they need to bring along enough energy to brake for a soft landing. In the absence of friction, things keep right on going until something comes along to stop them or turn them around. And it was this material stubbornness that Galileo recognized as inertia. The only reason that there aren't perpetual-motion machines is that this contrariness exists almost everywhere in the universe, even in space, where the average density is one atom for each cubic meter.[7]

Newton later amplified the idea to include explicitly the notion that anything which *didn't* keep right on going required some kind of force to stop it. He realized that the moon would keep right on going—literally flying off on a tangent—if some force didn't pull it toward the earth. That force, of course, is gravity. But what is the force behind the tendency of things to keep right on going? The force that makes the moon (or anything else) want to fly off on a tangent, or a yo-yo to fly off in a straight line when you whirl it about your head and then suddenly let it go? The truth is, nobody knows. "The motion to keep the planets going in a straight line has no known reason," writes Feynman. "The reason why things coast forever has never been found out. The law of inertia has no known origin." Good old inertia turns out to be one of the deepest mysteries of nature.

There is a clue, however, to the source of inertia. And that clue comes from the famous (if probably apocryphal) story of how Galileo climbed to the top of the Leaning Tower of Pisa and dropped two balls, a heavy one and a light one. The heavy ball and the light ball hit the ground at practically the same time. Even if the story is false, its lesson is true. Many science museums have exhibits consisting of vacuum tubes that allow visitors to watch as a feather and a coin fall to the bottom at the same rate. This doesn't happen in air, of course, because the air pushes back more on the relatively large surface of the feather than on the relatively small surface of the compact coin. But you can amaze your friends and neighbors (or at least your average six-year-old) by dropping a piece of paper and a small rock to see

---

[7]The pulsing of the electric and magnetic fields that make up light comes pretty close to perpetual motion, however. By the time it prints an image on your retina, light from a star may have been traveling for several million light-years.

which falls fastest. (The rock, obviously, but *not* because it is heavier, as you will see.) Now do the experiment again, but this time crumple the paper into a tight ball. Now the paper and rock should hit the ground simultaneously, proving that weight doesn't matter. It may be true that "the bigger they come, the harder they fall," but no one ever said anything about bigger things falling faster. This experiment probably wouldn't work from the tower of Pisa, because the thick pillow of air between the top of the tower and the ground would cushion the fall, but it works fairly well in your average living room, and perfectly well in a vacuum.

Why does it work? Heavy things and light things fall at the same rate in a vacuum for the simple reason that while gravity pulls harder on heavy things, heavy things also have more inertia—so they *resist* harder. This explains why a pendulum with a heavy weight at the end and a pendulum with a light weight at the end will swing at the same rate, and why a heavy object (like a satellite) will circle the earth at the same rate as a light object (like an astronaut inside it), making the astronaut weightless. For both the astronaut and the spaceship, the gravitational pull toward the earth is exactly balanced by the inertial tendency to fly outward. So if you drop your pencil while whirling around the earth in a space station, you don't have to worry about its falling to the floor. It will keep on orbiting, or "floating," exactly where it is.

Einstein thought it was a funny coincidence that gravity and inertia should balance each other so perfectly. In fact, he didn't swallow the coincidence at all. He looked for the reason behind the coincidence and concluded that gravity is not a force but rather the unseen curvature of the space we live in. This is the basis of his general relativity. And the "springboard" for this brilliant insight, writes Lincoln Barnett, was nothing other than "Newton's Law of Inertia which... states that 'every body continues in its state of rest, or of uniform motion in a straight line, unless it is compelled to change that state by forces impressed thereon.'" Even the equation that ignited the atom bomb ($E = mc^2$)—and in fact ignites every match and star and candle—first appeared in a paper entitled "Does the Inertia of a Body Depend on Its Energy Content?" So inertia is not only mysterious; it is also rich and profound.

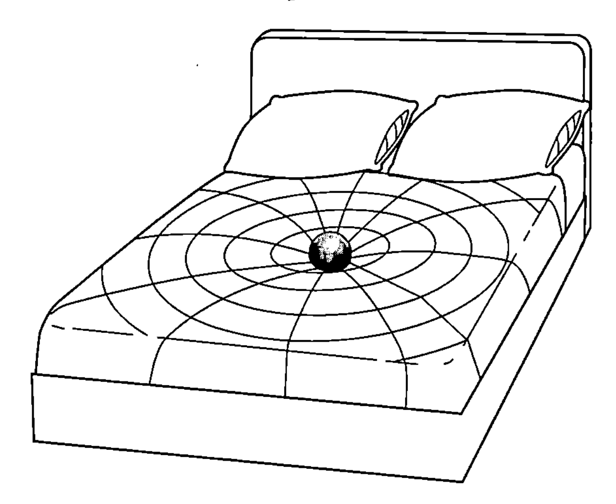

Einstein's general relativity describes the "force" of gravity as the unseen geometry of space. In this two-dimensional analogy, space-time near a massive star is curved in a way similar to the surface of a water bed when a heavy ball rests on it.

In fact, inertia may turn out to be nothing less grand than the cumulative pull of all the matter in the cosmos. If so, that's not just some lousy concrete pillar in a parking garage that thumped your bumper. It's the weight of all the stars in the universe. Indeed, we would be pushing off the stars every time we take a step, accelerate in a car, or blast a rocket into space.

This charming notion is known in physics circles as Mach's principle, after Austrian physicist and philosopher Ernst Mach, who proposed it more than a hundred years ago. It supposes that the resistance to change is due to the gravitational interaction of every bit of matter with every other bit of matter. Each particle, each star, each begonia, each automobile, is attracted to every other bit of matter by the force of gravity in a sticky, interlocking web. When you try to push a heavy sofa, you disturb the entire gravitationally entangled cosmos. No wonder it's so hard to move.

Until recently, no one was sure whether Mach's principle was consistent with Einstein's remarkably successful theory of gravity. According to Einstein, space and time are linked in a unified fabric that gets warped and otherwise distorted by very massive objects. Imagine an elephant sitting on a water bed and you get an idea of what happens to space-time in the presence of something really big, like a star or a black hole. If Mach's ideas are right, then massive rotating objects should drag space-time along with them as they move. In other words, if the elephant rolls over, the rubber surface of the bed gets dragged with him.

In late 1997, evidence for this dragging of space-time was finally seen around objects even larger than elephants—black holes and collapsed stars. If so, it's good news for Mach. The discovery, according to Italian physicist Luigi Stella, who claimed to see the signs of so-called "frame dragging" around collapsed stars, "shows us that [Einstein's theory of gravity] is in agreement with Mach's principle."

It's also good news for those of us seeking a sense of connection with the rest of the cosmos. Like a fly caught in a spiderweb, everything we do has an effect on the whole of the universe.

Hitting Back

> You can't be hit harder than you hit back. You can get only what you give. You cannot touch without being touched.
>
> —Paul Hewitt, *Conceptual Physics*

I still haven't answered my friend's question about "the force I feel when I stub my toe." In one sense, the force he feels is the combined gravitational pulls of all the matter in the universe—inertia. Except that inertia isn't considered a force. And the doorsill does more than merely *resist* a change in motion when you stub your toe. It actually pushes back. This is proved by the fact that stubbed toes are also squashed and painful toes. Something has exerted a force on them. In fact, the only reason you can walk at all is that the earth pushes forward on you when you push back on it, and the only reason you can drive a car is that the road pushes forward on the wheels as much as the wheels push back on the road. The earth doesn't move backward (very much) when you walk only because it's much more massive than you are.

This is the essence of Newton's famous law of action and reaction: "Whenever one body exerts a force on a second body, the second body exerts an equal and opposite force on the first." Forces always come in pairs. For every action, there is an equal and opposite reaction. It doesn't matter which is which, and often it's hard to tell which is which. The law of action and reaction can get rockets to the moon and even outside the solar system. If you fill up a balloon with air and let it go, it races around the room because the air being pushed out by the walls of the balloon is pushing back on the balloon, sending it flying. In the same way, a bullet pushed forward by a rifle kicks back with a power that can knock you to the ground.

This "reaction force" is more obvious in the absence of friction. When you use the force of your muscles to bang on a door, it is not obvious that the door bangs back on you, because friction keeps both you and the door more or less in place. If you banged on a door while you were wearing roller skates, however, your banging could easily propel you backward. And if two people on roller skates toss a ball back and forth, they'll drift farther and farther apart, because each time they throw the ball, the ball will throw them a little bit backward. The equal-and-opposite equation applies to all forces human and cosmic and mechanical. The earth pulls on an apple with a force equal and opposite to the pull of the apple on the earth, just as the equal and opposite pulls and pushes of light particles called photons account for electrical attractions and repulsions.

This cosmic game of tit for tat has the curious consequence that nothing can hit you back any harder than you hit it. Or as physics teacher Paul Hewitt likes to demonstrate in his classes, there is no way you can strike a piece of paper with an appreciable force. Hit it as hard as you like, and all you will ever feel is a slight tap. Since the sheet of paper cannot "hit back" with say, fifty pounds of force, it means that you cannot hit *it* with fifty pounds of force.

Of course, reaction implies a force, just as a resistance to change implies a force. But what kind of force is doing the reacting or resisting? We use the term *force* very loosely. Often, people say that they are forced to do something as a *reaction* to something done by somebody else. The force of habit is analogous to inertia. Physicists recognize a whole category of forces as "fictitious" or "pseudo" forces. Pseudoforces are those that appear to be forces in one frame of reference but not in another, just as one person may say he was *forced* to behave in a certain way, while to another person looking from a broader perspective it may appear that the first person is just following his normal behavior pattern. That is, whether or not you are actually *forced* to do something can be based on your point of view.

Say, for example, you are riding in a spacecraft that is accelerating for a fast trip to an another planet. Suddenly, you drop your wallet. If your spacecraft were just drifting around in orbit, then your wallet would float in space, but since the spacecraft is

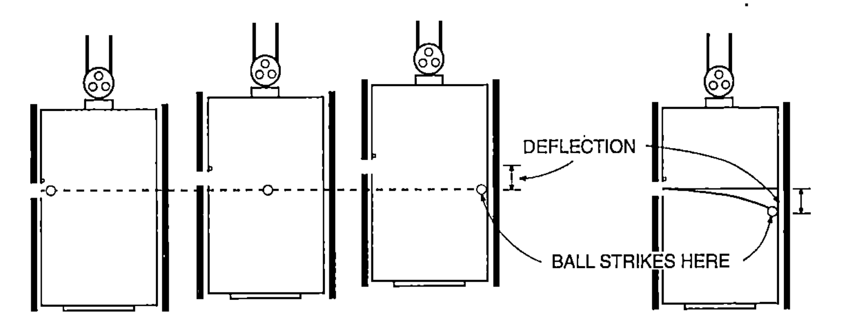

(Left) An outside observer sees a horizontally thrown ball enter the window and travel in a straight line, hitting the opposite wall. Because the elevator is rising, the ball hits the opposite wall at increasingly lower points. (Right) To someone inside the elevator, the ball appears to enter the window, then "fall" downward under the pull of gravity.

accelerating the floor soon overtakes the wallet and it appears to fall. From the point of view of someone sitting on, say, a passing moon, it would be clear that the wallet remained relatively stationary while the accelerating rocket raced to catch up with it. But to those inside the rocket, it appears that some outside force—gravity, or a magnet, perhaps—is attracting the wallet to the floor.

Either way, the wallet eventually hits the floor, or the floor hits the wallet. "Whether a force is fictitious or real becomes purely a problem of language," as physicist James Trefil points out. "It is a totally irrelevant question as far as any physical effect is concerned." The falling wallet is similar to the kind of "thought experiments" Einstein used to develop the idea of relativity—which doesn't mean that "everything is relative" but that no matter how you view the situation, the physical outcome is the same. Gravity, according to Einstein, is just this kind of pseudoforce. But that doesn't stop the moon from orbiting the earth or apples from falling to the ground.

For that matter, magnetism turns out to be an *entirely* relative force. That is, magnetism is always created by moving electric charge. The magnetism in iron magnets is caused by the spinning of countless electrons all twirling around in the same direction, just as the magnetic field of the earth is created (most likely) by the motion of electric currents deep inside its metallic core. Every time an electric current passes through a wire, it creates a magnetic field around the wire. Yet if you traveled along with an electron somehow, the magnetic force would seem to disappear, just as the motion of an airplane traveling 500 miles per hour seems to "disappear" when you're sitting in your seat watching an in-flight movie. Magnetism is a relative effect of electricity, just as the force of "wind" on your face on a still day is a relative effect of riding in a speedboat at 60 miles per hour. (Sailors, in fact, call this "apparent" wind—which is appropriately analogous to the physicists' pseudoforce.)

Four Fundamental Forces

O! O! O! you eight colorful guys
You won't let quarks materialize
You're tricky, but now we realize
You hold together our nucleis

—Anonymous, quoted in FRANK WILCZEK and BETSY DEVINE, *Longing for the Harmonies*

Reaction forces, relative forces, and even pseudoforces are always understandable in terms of at least one of the four so-called fundamental forces of nature. These forces are good old gravity (carried by as yet undiscovered particles called gravitons); the strong force (carried by gluons, which "glue" nuclear particles together); electromagnetism (carried by particles of light, or photons); and the weak force (carried by W and Z particles).

The last two forces have been united under a common parentage and are now sometimes known as the electroweak force.

The choice of four (or three) is hardly arbitrary. It has resulted from a long series of sometimes startling discoveries about how forces work and how they relate to one another. For a long time, for example, people thought that the force of gravity was balanced by an opposite force called levity, which caused things (like smoke) to rise. It took a great stroke of insight for Benjamin Franklin to realize that the static electricity that created small-scale sparks around the house was also the stuff of lightning, and it wasn't until the nineteenth century that electricity and magnetism were recognized as different aspects of the same thing. No wonder the search for unity among forces has such an attraction for physicists. It has been so successful and revealing in the past that it is only natural to assume that it will continue to unravel fundamental mysteries in the future.

Gravity

The most familiar force, of course, is gravity. It is the glue that keeps us stuck on the earth and pulls the elements of the earth in toward the center to form a compact sphere. It keeps not only our furniture from floating away, but also the air, the clouds, and even the moon. It makes the rain and baseballs fall. It is the force we all fight to push ourselves out of bed in the morning, the force we stand up against all day. We grow up in response to the way gravity pulls us down, so gravity determines our shape, whether we are trees or children or elephants. Gravity is the major force behind tides and weather and floods. It is even at the bottom of black holes. One of the first great "unifications" of forces was Newton's insight that the force that makes things fall on earth is the same stuff that controls the shape of things in the heavens. Gravity is truly universal. It is also insatiable: While an electric charge can find a mate and become neutralized, gravity never lets go. And while there is a limit to the number of electric charges that an atom will attract, there is no limit to the amount of matter gravity can pull toward a star.

Electricity

In 450 B.C., Empedocles of Agrigentum speculated that the earth was made of meal, cemented together with water. He was very nearly right, and almost completely right if you take meal to mean matter and substitute electricity for water. Electricity remains largely unnoticed until it flashes out at us during a lightning storm, but electrical forces are truly the "stuff" of matter.

When you stub your toe, it is really the outer electrons swarming about the atoms in your toe that collide with the outer electrons of the atoms in the wood, so electricity is "the force you feel" when you stub your toe. It is even electricity that makes it impossible to stub your toe in mush, because electricity accounts for all properties of matter: the hardness of wood, the transparency of glass, the glitter of gold. The interactions of those outer electrons buzzing around the atomic nuclei are responsible for everything from fire to thought, from cooking and digestion to taste and smell, from the solvency of water to the cleaning power of soap. Electricity is the force that makes sticky things stick together; it is behind the capillarity that pulls water up the trunks of tall trees and through the veins (and capillaries) of animals (including humans); it is even the source of friction.

Electricity becomes more impressive still when you consider that a moving electric charge produces still another force—magnetism. And that electricity and magnetism together form a continually alternating wave train that zips along through space at 186,000 miles per second, accounting for all radiation, including visible light, heat, microwaves, radio and television signals, X rays, and gamma rays.

Strong and Weak

The strong force and the weak force have remained hidden like genies inside the atomic bottle, springing into sight only recently. The strong force is sometimes called the nuclear force, for its realm is within the nucleus and its responsibility is nuclear reactions—most important, holding the constituents of the nucleus together. If there were no strong force, there would be no elements other than hydrogen with its single nuclear proton; there would be no planets, no life. The strong force is the force that fuels the nuclear reactor and the nuclear bomb, the sun and the stars. As physicists have seen farther into the core of the nucleus, they have discovered that the nuclear force is probably a kind of complicated effect (like chemistry) of a still more fundamental force called the color force, which has nothing to do with visible color. It is the color force that is carried by gluons, and that acts between quarks. Ultimately, it is gluons that hold quarks and therefore the nuclei of atoms together.

As for the weak force, suffice it to say that it is the force behind radioactivity and certain reactions in the sun. Radioactivity is of no small consequence, of course. It has kept the earth warm enough to support life, and the random mutations it causes have helped along the evolution of all species. Of late, the weak force has been revealed to have common roots with electromagnetism in a display of mathematical wizardry reminiscent of Maxwell's unification of electricity and magnetism more than a century earlier. It was this recent electroweak unification that pointed to the discovery of the W's and Z's—the weak-force carriers—at CERN.

A fast look at these forces tells you immediately why physicists are so interested in unity. They seem to have nothing whatever to do with each other. Gravity, for example, works only one way. It pulls everything *toward* everything else, which is one reason that so many things in the universe are round. It also explains why gravity is so noticeable. In truth, gravity is *trillions and trillions* of times weaker than electricity. We don't normally notice this powerful electric force directly, because in most of the universe it is balanced and therefore neutralized; that is, electricity, unlike gravity, both attracts and repels. All over the universe (and especially on the relatively calm and cool surface of the earth), bits of negative electric charge join with bits of positive electric charge to form bits of neutral matter. When you rub some of these charges off their neutral atoms (say, by shuffling across a carpeted room) and then allow them to get back together (say, by touching a doorknob), you can create small sparks. When the strong updrafts in huge thunderclouds rub raindrops together like your feet rubbing on the rug, huge numbers of electrons can be ripped off. When they get back together, they create the much larger sparks we know as lightning.

In other ways, gravity and electricity (or electromagnetism) are remarkably similar: Both decrease in strength as they spread out through space according to the same equation, and both can reach—theoretically, at least—to the ends of the universe. The mysterious color force, on the other hand, seems to *increase* infinitely as it spreads out from the vicinity of a quark. The farther two quarks drift away from each other, the more fiercely they are pulled back together. Thus, quarks are permanently trapped, and no free quark has ever been (or perhaps ever can be) found.

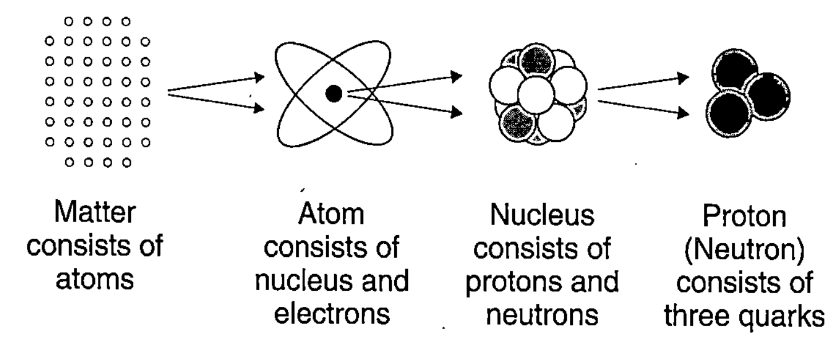

The different layers of the structure of matter.

Odder still, the little-understood color force seems to disappear altogether at very close ranges, leaving the quarks free to rattle around within a tightly closed bag. Just why the color force behaves the way it does remains the subject of intense research. Indeed, some physicists blame its odd behavior on the properties of the gluon "vacuum"—or empty space.

Forces not only act in very different ways, they also make themselves felt over very different ranges. While the gravity of the earth may contribute to the inertia of a distant star (and vice versa), the extremely parochial weak force exerts its influence over a distance a thousandth the diameter of a proton. Gravity is both the strongest and the weakest force in the universe, depending on the range and scale you consider. It is a major force in the lives of universes, people, and stars, but is virtually unfelt by atoms or even small spiders. The lives of lightweight things, from small plants to molecules, are ruled instead by chemical forces—surface tension, cohesion (stickiness), and capillarity—and all these forces are essentially electrical. The strong force is a major power within the nucleus but dwindles almost to nothing once it leaves the confines of that nucleus. It has strong but very short-range hooks that don't come into play until two nuclear particles get very close together.

Complicating the matter further, forces are picky. Different kinds of forces make their effects felt on particular kinds of things. Particles that carry the electric force can influence only electrified particles. Gluons carry color only between quarks and other gluons. The weak force is so specialized that it interacts only with left-handed particles and right-handed antiparticles. But gravity affects *everything*. The source of gravity is mass itself, and everything has mass, if only in the form of energy. (That is the meaning of Einstein's equation $E = mc^2$: energy equals mass times c [the speed of light] squared.) Therefore gravity can bend a light beam skimming by a star. This universal property of gravity is what allowed Einstein to see it as the grand geometry of all (curved) space.

Wrinkles in Time and Space

> In the physicist's recipe for the world the list of ingredients no longer included particles, but only a few kinds of fields.
>
> —STEVEN WEINBERG

In a sense, curved space is simply the shape of the gravitational force field that surrounds a piece of matter. A field is a kind of tension in space that extends out around a particle (or a planet) like the sticky spokes of a spiderweb, spreading its influence to all other particles that might come into its path, and even sucking in unwary particles like a vacuum cleaner. *Field* was the concept used to overcome the main objection to Newton's ideas about gravity. Newton never answered the question, How does gravity reach out and grab the apple or the moon? How does the force get from here to there? The concept of *field* eliminated the worry about "action at a distance" by temporarily eliminating the distance. The field spread the influence of the particle out into space itself.

At first, a force field was just an interesting way to look at how forces behave. That is, if you sprinkle iron filings near a magnet, they will line up in a shape that corresponds to the magnet's force field. In the same way, planets orbit the sun in a way that "lines them up" with the sun's gravitational field. The notion of field turns forces into geometry, making them into an integral part of the landscape itself. Say, for example, you threw a ball down an invisible pipe that was bent in several places. You could say that the ball was "forced" to follow the bend by the pressure of the walls of the pipe, or you could say that the ball naturally traveled in a bendy fashion because the space around it curved. Large-scale force fields (like gravity) are said to curve space, while small-scale forces (like electromagnetism) are sometimes said to "wrinkle" it. (Actually, they wrinkle the electromagnetic field.) Either way, a force field is a useful mental and mathematical tool for visualizing how forces operate. It is a description of the effect of a force which tells the strength and direction of that force at every point in space.

It turns out, however, that force fields are much more than that. Force fields exist *on their own,* independent of the particles responsible for creating them. A planet or an electron is the source of a force that creates a tension in the space around it. But that force—that tension—can exist in the absence of the planet or electron. When a wiggling electron in the sun sets up a local wrinkle in its surrounding electric field, that wrinkle speeds along at 186,000 miles per second and almost eight minutes later sets up a wrinkle in an electric field somewhere on earth, where it might be detected as "light." That wiggle takes time to get from there to here. If the sun went out during the eight minutes it took the light particle to travel to the earth, we would still be able to see it. When a star explodes, the gravitational effects linger years after the event. Force fields have lives of their own.

The idea that a force field could be a separate entity was an important first step toward the notion of forces as particles. The second step was quantum mechanics, which grew out of the finding that everything, including the energy of force fields, is quantized—that is, comes in clumps. Therefore a force particle, like a photon, is really a small clump of an electromagnetic field that travels from place to place at the speed of light, carrying its quantized parcel of energy and momentum with it. Force fields and force particles may seem like very different things, but most of the differences lie in our dearth of proper imagery.

Nevertheless, the imagery of force particles has become

deeply rooted in physics, and especially in popular writing about it. You'll often read about particles being "exchanged" between other particles, like the ball "exchanged" between two people on Rollerblades. The force of the exchange, says the imagery, is what drives the particles apart. This imagery doesn't explain attraction, however. In this case, two particles are drawn together by "sharing" the same force particle, much as two people can be drawn together by sharing an umbrella. Exchanges of words can be used either to attract or repel. Many chemical bonds are the result of a sharing of electrons among atoms.

If forces can be particles, however, is there any difference between matter and the pulls and pushes that make it come and go? Is there any difference between "stuff" and "influences"? Can you separate the actions from the actors—the things people (or particles) do from the people or particles themselves? Einstein speculated that any divisions between the two were "somewhat artificial." Today physicists describe everything in terms of fields.

There is one important difference between matter and forces, and it is this subtle difference that causes pain when you stub your toe. Matter particles (like protons, neutrons, and electrons) obey what is known as the Pauli exclusion principle, named after Austrian-born physicist Wolfgang Pauli. The notion of atomic electron shells rests on the Pauli principle, because it says that no two electrons can occupy the same state. If all the spaces or states in shell number two, say, are full, then the next added electron must occupy a space in the next available shell. The Pauli principle explains why matter is not compressible, so it is really a "principle" that makes things solid and also the real force you feel when you stub your toe.

This same principle (also known as electron pressure) is what keeps stars from collapsing. When a star is so massive that its gravity overcomes electron pressure, it literally collapses: The electrons compress into the nucleus, joining the protons to form neutrons and thus making a neutron star. If even the nuclear forces aren't strong enough to hold back gravity, the star keeps on collapsing—in theory, at least, forming what is known as a black hole.

Force particles, on the other hand, do not obey the Pauli exclusion principle. You wouldn't stub your toe if you walked into a light beam. Force particles are described by a different set of statistics, called the Bose-Einstein statistics, which is why force particles are called bosons. Most matter particles are described by what are known as Fermi statistics, after Italian physicist Enrico Fermi, and so are called fermions.

In 1995, physicists muddled even this division by producing what was widely hailed as a new kind of matter, a single "superatom" made of matter particles that condense like force particles into a single entity. Created at the University of Colorado in a carrot-size tube containing the coldest spot in the universe, the so-called Bose-Einstein condensate lasted for about 20 seconds at about 170 billionths of a degree above absolute zero, or minus 459 degrees F.

It's not surprising that this curious state was reached at the extremes of temperature, just as Einstein's relativity comes into play only at extreme speeds. Extremes are where the laws of physics break down, opening cracks so that scientists can peer into uncharted realms. Today physicists at several particle accelerators around the world are hot on the heels of the so-called quark-gluon plasma, a state of matter so hot that quarks and gluons condense together in a primordial cosmic soup where force and matter are indistinct. Matter came into being, physicists think, under just such conditions.

Like light skimming around an object and casting a fuzzy shadow, the line between force and matter tends to blur at the edges.

CHAPTER SEVEN

Quantum Leaps

> That, in a nutshell, is the mystery of the quantum: When an electron is observed, it is a particle, but between observations its map of potentiality spreads out like a wave. Compared to the electron, even a platypus is banal.
>
> —Hans Christian Von Baeyer, *Taming the Atom*

The introduction of quantum theory in the early 1920s marked one of the greatest revolutions in all of physical science. It could not (cannot) adequately be described in metaphors borrowed from our previous view of reality, because many of those metaphors no longer apply. This inability to imagine quantum goings-on led to the popular perception that the realm of the inner atom is fuzzy, elusive, murky, and uncertain. On the contrary, most physicists would agree that what quantum theory has brought to science is exactly the opposite—concreteness and clarity.

What is quantum mechanics? In the simplest sense, it's the mechanics of quantized things. Mechanics is the explanation of the way things work in terms of energy and forces and motion. Before the turn of the century, the way things worked was pretty well explained by Newtonian mechanics, also called "classical" mechanics. Everyone knows about typical Newtonian classical systems: billiard balls colliding and veering off on precisely predetermined paths with precisely measurable amounts of energy and momentum; planetary systems orbiting central suns according to the laws of gravity. Once Ernest Rutherford had "seen" the atomic nucleus, the atom itself was considered just such a planetary system: Electrons orbited the nucleus like planets orbiting the sun. And atomic particles behaved, of course, like billiard balls—albeit very *small* billiard balls.

But the central feature of Newtonian mechanics was that everything was continuous: things flowed smoothly through space; energy could come in an infinite range of amounts; light undulated in continuous waves; there was no minimum amount of anything.

Quantum mechanics changed all that. Now energy, light, force, and motion (among other things) are known to be *quantized.* You cannot have just any old amount; you can have only multiples of certain minimum quantities. Quantum mechanics meant that all the *qualities* of subatomic things—and by extension, of all things—were precisely *quantifiable.* In a sense, this made things nice and neat and even more "scientific" in the popular sense. But it also meant that the smooth continuity of the Newtonian universe was lost. Now nature is revealed to be somewhat jerky or grainy, jumping from one quantum amount to the other, never traversing the area in between.

Moreover, this leads to an uneasy uncertainty about what is going on *between* those quantum states, *during* those quantum leaps. In fact, it turns out that there is no way of knowing the precise state of things between quantum states. In human terms at least, there *is* no transition between quantum states. You can have either one or two or three units of energy or momentum or light or force or matter or whatever, but there is *no such thing* as one and one-half or two and three-quarters units. Everything in the quantum mechanical universe (which is our universe, of course) happens in quantum leaps.

In other words, quantum mechanics looks murky or clear at

least in part depending on which of the two complementary aspects you focus on. If you focus on aspects such as the notorious Heisenberg uncertainty principle, which explains why any precise measurement of one aspect of a quantum particle (say, position) introduces uncertainty about another (say, motion), then quantum mechanics looks murky. If you focus on the definiteness of quantum units, it looks concrete.

Victor Weisskopf goes as far as to say that the uncertainty principle should really be called the definiteness principle. His reasons are compelling: Consider a typical Newtonian (pre–quantum mechanical) system—nine planets orbiting a central sun. The laws of classical physics allowed for a tremendous amount of flexibility. All the law of gravity requires is that the nine planets orbit the sun in more or less elliptical paths; nothing says that the orbit of the earth or of any other planet must be "just so." In fact, depending upon the initial conditions present at the formation of the solar system, almost any elliptical orbit would do. And if someday a passing star should come along and push our world out of its place in the sun with the force of its gravity, our orbit would be irrevocably changed. There is nothing whatever special about our present position that would cause us to remain here—or to spring back to our original orbit once we had been pushed out of it. Since planets orbiting a star can fall into almost any possible elliptical path, other solar systems surrounding other stars are likely to be found (and indeed have been found) in a great variety of arrangements.

Contrast this with an atomic system, which was pictured as just such a miniature solar system. The central nucleus attracts the orbiting electrons just as the sun attracts the orbiting planets. But here the similarity ends. In solar systems, an infinite number of stable configurations is possible; in atomic systems, there are only about a hundred—each corresponding to one of the known elements. If the system contains one nucleus (one sun) and one electron (one planet), it takes a single, unchanging form—the hydrogen atom. All hydrogen atoms are exactly alike. And if a bunch of hydrogen atoms get bounced around by other hydrogen atoms or by anything else, they automatically spring back to their original shape.

"Before we got quantum theory," says Weisskopf, "our understanding of nature did not correspond at all to one of the most obvious characteristics of nature—namely, the definite and specific properties of things. Steam is always steam, wherever you find it. Rock is always rock. Air is always air. Two pieces of gold mined at two different locations cannot be distinguished." Even two pieces of gold found in different galaxies would have identical properties.

It should be highly improbable to find two atoms exactly alike, just as it would be highly improbable to find two solar systems exactly alike. But obviously it's not. A carbon atom enters your body as part of a sandwich and emerges hours or days or perhaps even years later, after taking part in countless chemical reactions, still as a carbon atom. The particular combination of protons, neutrons, and electrons that makes up a carbon atom can *only* be arranged "just so." Quantum theory brought just this exactness into our understanding of atoms. If anything was uncertain, it was *classical* systems.

Too many good books have already been written on quantum theory for me to even attempt to go into the details of its development in this short chapter. Suffice it to say that it was Niels Bohr in the early 1920s who came up with a way to understand the stability and exactness of atoms using the analogy of standing waves.[8] Take a jump rope, or a violin string, secured at both ends. If you pump energy into it and set it swinging, it

[8]Of course, the development of quantum theory did not take place in quantum leaps but happened over a period of many years and involved many different physicists.

can vibrate only in a certain number of ways, taking a few predetermined shapes. A violin string vibrates in its fundamental frequency, or in twice, three times, or four times that frequency—in other words, in its characteristic harmonics. It cannot vibrate in two and one-half times that frequency. If you imagine an electron "wave" confined to an atom in much the same way, you can easily understand how it would be forced to assume only a certain number of predetermined vibrational states. Atoms—unlike solar systems—are innately stable and consistent, because electrons cannot take their places around the nucleus just anywhere. Every hydrogen atom in the universe strikes the same chord of frequencies. (Weisskopf told me that he once tried to play the chord for hydrogen on his piano: "It sounds terrible," he said, "but then it's not music for our ears.")

There are other analogies, of course, though none quite as useful as the image of standing waves, since electrons in fact do behave like waves. But just for argument's sake, one could imagine the electron in the atom as a child jumping up stairs. Depending on the amount of energy she has, she can jump up to the second stair, or the fourth, or the sixth, but she cannot land safely or remain stable at step two and one-half, or three and one-quarter. She needs a minimum amount of energy before she can attain the next step, or state. If she doesn't have quite enough energy to make it to step four, she will remain at step three. And atoms will not absorb radiation unless the energy they receive contains the minimum to make the next "quantum leap."

Or you could think of the girl jumping *down* the stairs. In this case, she gives off energy to the floor as she jumps to a lower state. But the energy comes in clumps. A jump from step four to step two gives off two "steps" worth of energy, and so on.

An electron jumping to a lower state gives off its energy in the form of light. A jump from "orbit" or step five to orbit three might radiate clumps of red light; a jump from six to two might radiate a more energetic (higher-frequency) blue light; a jump from orbit two to orbit one (the ground state) might give off very low-energy radio waves; a jump from orbit eight to the ground state might give off a highly energetic X ray. There is a similar series of even more energetic quantum states within the nucleus that account for radiation such as gamma rays.

Each color emitted by an atom, in other words, corresponds to a specific change in energy state—or quantum leap. Together, the colors—the spectrum emitted by each element—are as unique as a signature. And the spectrum of sodium or mercury or hydrogen looks exactly the same no matter where it's found—even on the farthest stars.[9] These quantum mechanical fingerprints can reveal the chemical makeup of distant stars, or clues left at a crime scene.

Hot metals (like filaments in lightbulbs) glow with a continuous spectrum of colors, and not with the sharp spectral lines characteristic of quantum leaps, because they emit light in a different way from gases. George Gamow described the spectrum of an excited gas as the harmonics of a single instrument, but atoms in a solid are so closely packed together that it is as if you threw all the instruments of a symphony orchestra into a bag and shook them up. The spectral colors that shine from hot metals come not from electrons bound to atoms but from free electrons that are, well, free to vibrate more or less at random.[10] They do

---

[9]Actually, the spectrum doesn't look *exactly* the same on stars, but the differences tell astronomers things such as how fast the star is moving and how far away it is.

[10]The frequency of vibration of light—that is, how rapidly the light waves oscillate, or move up and down—determines its color. Higher frequencies produce shorter waves; in fact, if you shake or vibrate a rope, you can see that faster vibrations make shorter waves. It also takes more *energy* to make smaller, more rapidly vibrating waves. With light, too, short waves (fast vibrations) correspond to high energies. In the electromagnetic spectrum, the most slowly vibrating "light" falls into what we call the radio frequencies; these waves can be as large as mountains. Visible light waves average about 1/50,000 of an inch long. X rays are as small as atoms.

not have to jump from step four to one, or six to two. They can give off a whole range of frequencies of light. For this reason, anything that gets hot enough burns with the same colors; it does not matter what is on fire.

Sometimes so much energy can be pumped into an atom that the electrons dislodge entirely. It ceases to be an ordinary atom of carbon or neon or oxygen and instead becomes an ion—an electrically charged "part" of an atom. If *all* the electrons are knocked off, then you are left with a special kind of gas known as a plasma—an amorphous mixture of nuclei and electrons. There are no atomic quantum states in plasmas. (There are no *atoms* in plasmas.) Yet most of the matter in the universe exists in this highly energetic form. Stars are essentially balls of plasma; the spectral lines that tell what stars are made of come from the cooler atoms on the star's surface.

At earthly temperatures, however, the quantum reigns. "Ultimately," says Weisskopf, "all the regularities of form and structure that we see in nature, ranging from the hexagonal shape of a snowflake to the intricate symmetries of living forms in flowers and animals, are based upon the symmetries of these atomic patterns." The fact that you inherit stable genes from your parents is based on the inherent stability of the quantum states of the molecules that make up DNA. The hardness of wood, the softness of tissue paper, the scent of flowers wafting in through the window—all are consequences of the quantum states of atoms.

As it turns out, everything in the subatomic world is quantized: not only energy and light but also matter, "action" (energy times time), momentum, spin, electric charge, and all the other exotic qualities of subatomic things—like "strangeness" and "charm." You cannot even conceive of an action, or a motion, or a bit of matter, smaller than this minimum. An atom absorbing any one of these qualities has to swallow it whole or not at all; it spits it out in quantum clumps. This means that the very stuff of the universe cannot be smoothed out beyond a certain point; it has a texture; it is grainy, or lumpy.

In truth, this seems odd and unpleasant only until you stop to think about it. After all, many things in everyday life are lumpy, too—auto transmissions, for example. As Gamow points out, the gearbox on your car is something like the quantum states of atoms: "One can put it in low, in second, or in top gear, but not in between." (Of course, you can physically put it in between, but that doesn't have much meaning as far as the motions of your car are concerned.) Another familiar phenomenon that comes in clumps is people. Richard Feynman uses the familiar example of the "average American family" consisting of 2 adults and 2.2 children, or whatever. Everyone knows it's a joke, because everyone knows that children, like quanta, are units. Some would even argue that states of mind—even leaps of the imagination—are similarly quantized. There is rarely a continuous transition from one idea or state to the next. (As Einstein said, "There comes a point where the mind takes a leap—call it intuition or what you will—and comes out on a higher plane of knowledge.") Cultures, perceptions, beliefs, and even phases of life often seem as discrete as individual quantum states, which is why we sometimes feel truly transformed when we move from one to another. This is not entirely surprising, because, of course, even the brain is quantized: An impulse is transmitted by a nerve cell as a whole or not at all.

Most of this is metaphor, of course. But the metaphors come easily enough to suggest that a quantum is not a completely unfamiliar concept. In essence, it means only that some things need to be considered whole: children, snowflakes, atomic states, and also memories, experiences, poems, paintings, and a whole host of other things. It also embodies an irreducible on/off, yes/no quality—something that should be intimately familiar to anyone weaned in the computer age. A person, for example, can have a whole range of degrees of a quality such as charm, but when it comes to atoms, either you've got it or you don't. The same is true of energy, spin, "strangeness," electric charge, and so on.

On a less metaphorical level, almost anything that has to do with waves has these very specific properties reminiscent of quantum states. Not only a violin string but also the air inside flutes and organ pipes vibrates in fundamental frequencies or exact multiples thereof, each time taking a quantum leap to the next harmonic. Even a single wave is a quantum, in that it has to be considered whole. It makes no more sense to talk about a third of a wave than it does to talk about a third of a child.

In another context, Stephen Jay Gould argues that our fear of quantum leaping comes from "a deeply rooted bias of Western thought that predisposes us to look for continuity and gradual change: *Natura non facit saltum* (nature does not make leaps), as the older naturalists proclaimed." Gould is one of a new breed of naturalists who are convinced that even evolution probably occurred in something similar to quantum leaps. "Change is more often a rapid transition between stable states than a continuous transformation at slow and steady rates," he writes, in terms that could also apply to atoms. Evidence accumulated from the fossil record seems to deny older theories that evolution proceeded gradually through small, continuous adaptations; rather, species seem to appear, stick around for a while virtually unchanged, then disappear again. Gould was surprised to find that many Soviet scientists already shared this perception—at least in part, he speculates, because of their training in the dialectical laws, laws that suggest "that change occurs in large leaps following a slow accumulation of stresses that a system resists until it reaches the breaking point. Heat water, and it eventually boils. Oppress the workers more and more and bring on the revolution."

Quality and Quantity

> Dust, sand, pebbles, rocks, and boulders are all made of the same material. They just contain different quantities of the same stuff. But we view them as qualitatively different. So in that sense, quantity is quality.
>
> —Philip Morrison

This notion that merely having *more* of something (or less of it) can change the nature of things *qualitatively* is at the heart of quantum mechanics. But it also can be one of the hardest aspects to accept. Physicists tend to give people the creeps when they start describing the fundamental building blocks of nature in terms of quantum "numbers." How can the addition or subtraction of single units of spin or electric charge or other "quantities" make the difference between neon and sodium? Or ultimately apples and oranges? The immediate explanation, of course, is that units of electric charge, for example, determine the chemical properties of things, but there is a much more interesting and fundamental relationship.

One of the first people to recognize the intimate connection between quality and quantity was the sixth-century B.C. Greek philosopher Pythagoras. He discovered that the pitch of a note depends on the length of the string producing it, and that pleasing intervals are those corresponding to simple mathematical ratios: 2:1 producing an octave; 3:2, a fifth; 4:3, a fourth; and so on. The quality of musical tones—like the qualities of elements—is based on the numerical relations of their parts.

Today the evidence that quantity affects quality is everywhere. The quantity of money or education you have may or may not make a difference in the quality of your life, but the quantity of pollution, noise, and crime in your neighborhood almost certainly does. Rapid increases in the overall quantity of otherwise beneficial things (cars, plastics, houses—even people) can turn them into qualitative disasters. Medicine taken in too large doses can be poisonous. And bigger (nuclear) bombs do not merely produce bigger wars: The difference between killing millions of people and wiping out civilization is qualitative, not quantitative.

One of the most impressive qualitative changes produced by sheer increase in quantity is the human brain. "When people evolved from the animal kingdom," says physicist Weisskopf, "something new must have happened. We contend that this new element is based solely upon a quantitative difference in the nervous system. By an increase in this system, nature established a new type of evolution that broke, and will break, all rules established in the previous evolutionary periods." Naturalist Gould is similarly impressed: "Perhaps the most amazing thing of all is a general property of complex systems, our brain prominent among them—their capacity to translate merely quantitative changes in structure into wondrously different qualities of function."

It takes a large quantity of atoms to make an organic molecule, and a large quantity of organic molecules to make a person, and a large quantity of people to make a crowd or a country. But in each case, it's clear that the difference is a lot more than quantitative.

This ability of quantitative changes to produce dramatic qualitative differences also becomes dramatic at extreme temperatures. As mentioned before, matter at very high temperatures actually falls apart; electrons leave the atomic nuclei and form an undifferentiated plasma, the stuff of stars. Plasma is nothing like ordinary earthly matter. For one thing, its electrically charged particles are not bound together in atoms but behave independently. For another, it does not come in well-defined varieties, such as silicon, oxygen, and lead. At higher temperatures still, an even more extreme state of matter is created in which atomic nuclei come apart into their constituent protons and neutrons. At yet higher temperatures (perhaps at the origins of the universe), even protons and neutrons disintegrate into a whole host of exotic species. If the esoteric antiparticles and mesons and J/psi's that pop up in high-energy accelerators seem strange to us, it is because they represent a *qualitatively* different kind of matter than we are used to.

As temperatures go down, matter again transforms into a whole series of qualitatively different states. The transition from gas to liquid takes place when a certain quantitative difference in temperature is reached; solids and crystals form at lower temperatures still. At extremely cold temperatures, there exists yet another state of matter, as exotic as its superhot counterparts. Certain materials at supercold temperatures become "superconducting": They can carry an electric current forever without resistance. Supercold helium becomes a "superfluid" that flows up and out of bottles, and down through the bottoms of ceramic containers. Matter becomes so ordered and so quiet that subtle quantum effects make themselves visible. When all the random motion that makes up heat is taken away, one can hear the inner whispers of atoms, "like the murmurings of a seashell," as one physicist put it.

Similarly, quantitative changes in size have huge qualitative effects. A billiard ball the size of a star does not behave like a billiard ball the size of an atom. A ball of matter the size of an ordinary billiard ball cannot collapse under its own weight, but a ball of matter the size of a star easily can. A ball of matter the size of an atom doesn't behave like a ball of matter at all, but more like a wave. Curved space is primarily a property of huge things; quantum effects apply only to tiny things. On the scale of an atom, gravity is insignificant; on the scale of the universe, it is the most significant force there is. Indeed, one of the most pressing preoccupations of present-day physics is to find the connections between the behavior of the very large and the very small—the so-called quantization of gravity.

Sir James Jeans pointed out that while philosophers generally think in terms of *qualities,* physicists normally describe things according to their *quantities:* "The philosophical lecturer may be telling his audience that a lump of sugar possesses the qualities of hardness, whiteness, and sweetness, while his colleague in the science room next door may be explaining coefficients of rigidity, of reflection of light and hydrogen-ion concentration—measures of the degree to which the qualities of hardness, whiteness, and sweetness are possessed."

And yet it is difficult to draw a clear difference between the two. For if hydrogen-ion concentration is a measure of sweetness, and reflection of light determines whiteness, then quality *is* quantity in a very real sense.

For that matter, talking about quantum mechanics at all involves taking a kind of quantum leap into a new dimension, where almost everything is qualitatively different—and mainly because the things we are talking about are quantitatively so much smaller. Einstein knew that going to extremes of size or speed could lead to qualitatively surprising results. Throughout the history of science, the most fruitful areas for exploration have hovered at the extremes and fringes—the outer limits of hot and cold, fast and slow, big and small, few and many. Or as a physicist friend once told me, "Everything in the middle is engineering."

So if quantum mechanics seems weird to us, perhaps that's only natural. Atoms, from our perspective, are unimaginably small. It should not be surprising that they do not behave as we do.

CHAPTER EIGHT

Relatively Speaking

> It now is nearly a full century since Einstein destroyed Newton's concept of space and time as absolute, and began laying the foundation for his own legacy. Over the intervening century, Einstein's legacy has grown to include, among many other things, a warpage of spacetime and a set of exotic objects made wholly and solely from that warpage: black holes, gravitational waves, singularities (clothed and naked), wormholes, and time machines. At one epoch in history or another, physicists have regarded each of these objects as outrageous.
>
> —KIP THORNE, *Black Holes and Time Warps: Einstein's Outrageous Legacy*

ALMOST FROM the beginning, relativity has been considered as much a philosophy as a physical theory. Einstein himself guessed that more clergymen were interested in it than physicists. Perhaps this had to do with the fact that relativity touches deep roots in culture, history, and religion: Our worldview depends very much on cosmic notions of space and time, which in turn determine how people fit into the scheme of things. Yet whatever the reasons, attempts to popularize and philosophize about relativity have almost always got it wrong. What's filtered down as the popular perception of relativity is the phrase "Everything is relative." In fact, the implications of Einstein's ideas are almost exactly the opposite.

What many people mistake for the theory of relativity goes

back at least as far as Galileo, and it has to do primarily with relative motion. That is, if you were riding inside the closed cabin of a steadily moving ship, you could not tell whether you were truly moving or not. You could perform all manner of experiments—throw balls, swing pendulums, try to balance on your toes—and everything in the cabin would behave exactly the same whether you were moving or standing still. The reason is that while you would be moving relative to the world outside the ship, you would not be moving relative to the world inside it—just as when you are relaxing on your front stoop you are not moving relative to the rapidly spinning earth, and thus you perceive no motion. Motion is relative because it depends on your point of view. What's moving in one frame of reference is not necessarily moving in another.

Say you asked a person standing on a nearby dock to tell you whether or not you were moving. The person could answer, "Yes, of course you are moving, you dummy. Can't you see the dock moving past you?" To which you could reply, "But how do I know that it's me and not the dock that's moving?" (And of course the dock *is* moving along with the motion of the earth, but that is another matter.) Or else the person on the dock could answer, "You are moving relative to my dock, but as long as you stand still in the cabin there, you are not moving relative to your ship." Or else the person could shrug her shoulders and say, "Well, everything is relative, you know."

The question would be much more difficult if you wanted to find out whether or not the *earth* was moving by asking such an outside observer. For where would you put her "dock"? How could you find her a frame of reference where she would be standing still? The earth moves relative to the sun, but the whole solar system moves relative to the galaxy, and the galaxy moves relative to the rest of the universe, and by all indications the universe is moving (in some sense) too. But relative to what?

"One of the great clichés about Einstein's theory is that it shows that everything is relative," wrote James R. Newman in *Science and Sensibility.* "The statement that everything is relative is as meaningful as the statement that everything is bigger.... If everything were relative, there would be nothing for it to be relative to."

Einstein took Galileo's relativity and turned it on its head so that it came out as what Victor Weisskopf likes to call the theory of absolutism. Einstein looked inside the cabin of the moving ship (except it was a light beam) and was profoundly

This photograph shows the pattern made by a ball moving back and forth in a straight line between two rotating turntables. If you moved along with the ball in the same way as you move along with the spinning earth, the complex pattern would vanish. Motion is relative because it depends on your point of view.

impressed *not* that everything is relative but that the laws of nature stay exactly the same. Jump up, and gravity pulls you down the same way whether you are moving or standing still. Throw a ball, and it follows the same path. Water flows, clocks tick, raindrops fall, and electricity attracts and repels just the same whether you are moving or at rest. "Relativity established that the laws of nature are absolute and do not depend on the motion of the system," says Weisskopf. "It's *because* they are absolute that you cannot tell whether you are moving or at rest."

Of course, certain things that people thought were absolute have to be sacrificed in order to achieve absolutism in the laws of nature. Space and time become relative, but in the scheme of things space and time turn out not to be very important, or at least not as fundamental as many other things.

What is much more important is that the conventional concepts of physics embodied in Newton's laws simply don't work at very high speeds or under conditions of extreme gravity or in many other situations. Newton's laws do not hold true in all frames of reference, so the laws of nature depend on whether you are moving and what system you are in. The laws of nature depend on your point of view. Now, that's a theory of relativity!

Galileo and Newton realized that motion was relative, but insisted that space and time were absolute. Einstein saw that space and time—and energy and mass, for that matter—were also relative, but that this was a peripheral effect of the absolute nature of other fundamental constants, among them the speed of light. For all the odd effects of relativity flow from the even odder fact that the speed of light is always absolute—an absolutely unchanging 186,000 miles per second from any point of view, from any frame of reference, moving or not. The only reason "everything is relative" is that the speed of light and the laws of nature are not. And since light itself is nothing but the motion of electric and magnetic fields relative to each other—forces that are behind everything from the nature of matter to all human processes including perception—Einstein clearly hit upon a fundamental "absolute" frame in which to construct his relative universe.

The Relativity of Time

> Absolute, true, and mathematical time, of itself, and from its own nature, flows equably without relation to anything external.
>
> —Isaac Newton, *Principia*

Almost everyone accepts Newtonian time unthinkingly. Yet the more you think about it, the less sense the idea of time ticking away isolated from everything else in the universe makes. Every notion of time you can imagine is intimately connected to a concrete physical event—the swing of a pendulum, the orbit of the earth, the vibrations of a quartz crystal, the quantum leaping of atoms, the motions of magnetic and electric fields, the lives of suns, and so on. Without such events, what would time consist of? You can't have time in a void, because there would be nothing, so to speak, to "tick." Time makes sense only when it's connected to things.

The time interval we call a year marks a single revolution of the earth around the sun; the day is a single spin of the earth around its axis; and long before humankind appeared, the month probably matched the orbit of the moon around the earth. It does so no longer simply because the moon's orbit is continually changing as the moon moves farther away. Clearly, none of these astronomical measures of time could ever be considered absolute, if only because they *are* constantly changing. Some 500 million years ago, our day was only twenty and a half hours long. And just to remove ourselves from our parochial earth perspective for a moment, consider the concept of time on a planet like

Mercury, where the day is longer than the year. It's a difficult idea to get used to, because we are so accustomed to thinking of "days" as the natural division of "years" into 365 parts. We forget that the day, like the year, is a happening, not an empty interval of some amorphous quantity we call time.

Other standards of time have no natural origins at all. What is the natural origin, for example, of the seven-day week? Some people say it derived from Genesis (on the seventh day, we rest). Others link it to the seven notes on the familiar Western musical scale which corresponded to the seven classical Pythagorean heavenly bodies (sun, moon, and five visible planets), which in turn played the music of the spheres. Various cultures at various times have tried out eight-day weeks, five-day weeks, and even a decimal ten-day week.

The hour is actually a quite recent addition. Until the fourteenth century, days were divided into much less regular intervals of morningtide, noontide, and eventide. The first hours were flexible: They varied from summer to winter, and from daylight to darkness. Well up into the Middle Ages, each day (dawn to dusk) and night (dusk to dawn) was divided into twelve equal parts. This meant that the hours of a summer day lasted much longer than the hours of that same summer night. Winter daylight hours were correspondingly shorter, and winter night hours correspondingly longer. Even regular hours would not do, however, when the Industrial Revolution made it necessary for the trains to run on time and the workers to arrive for the five o'clock shift, so minute hands sprouted on clocks like so many offspring of a new age.

But all these concepts of time are modern inventions. As Arthur Koestler points out, less than fifteen generations ago "time... was simply the duration of an event. Nobody in his senses would have said that things move *through* or *in* space or time—how can a thing move in or through an attribute of itself, how can the concrete move through the abstract?" The idea that time was an attribute of things—like length, width, and depth—seems much closer to Einstein's four-dimensional space-time than to the abstract clock time we carry around with us today.

Indeed, depending on what means you use to measure time, time itself changes character completely. The atomic year of hydrogen, at 10^{-16} seconds (one divided by the number ten with sixteen zeros after it) is infinitely short compared to an earth year. (An atomic year is the time it would take an electron to "orbit" the nucleus if the atom were a miniature solar system.) Yet the atomic year itself is infinitely long compared to the fundamental time units pertaining to nuclear particles, which are millions of times shorter. To a particle in the nucleus, an electron would seem almost stationary, just as the "fixed" stars seem stationary to us.

In geological time, on the other hand, an instant can be 10 million years, because 10 million years is but 1/450 of the earth's history. A thousand years is an interval so short that geologists can rarely resolve it. It's all but undetectable, and so is treated as a passing moment. "With our short memory," writes Loren Eiseley, "we accept the present climate as normal. It is as though a man with a huge volume of a thousand pages before him—in reality, the pages of earth time—should read the final sentence on the last page and pronounce it history."

The universe is atick with all kinds of timekeeping devices, all clocking different kinds of time. Radioactivity provides a natural clock that resides within atoms. Certain atoms exist in unstable forms that after a while decay into more stable forms. An ordinary carbon nucleus, for example, contains six protons and six neutrons and is known as carbon-12. But there is another form, or isotope, of carbon—carbon-14—whose nucleus contains six protons and *eight* neutrons. During radioactive decay, one of those extra neutrons emits a negatively charged electron and becomes a positively charged proton. And faster than you can say "alchemy," the unstable carbon nucleus has changed into a stable nitrogen nucleus with seven protons and seven neutrons.

What's even more fascinating about these transformations, however, is their *timing*. Every 5,700 years, exactly half a given number of carbon-14 nuclei will decay into nitrogen-14 nuclei. If you started with 10 trillion carbon-14 atoms, you would have 5 trillion carbon-14 atoms 5,700 years later. And 5,700 years after that, you would have exactly 2.5 trillion carbon-14 atoms.

The half-lives of atoms are remarkably accurate timekeepers because nothing outside the atom can affect them. They are immune to external influences. And unlike astronomical measures, they never change. Still, they provide a strange kind of clock that ticks only for large numbers of atoms. It does not work at all for a single atom. It is a clock based on the statistics of probability.

And then, of course, there are all manner of biological clocks. All living things need to tell time to survive, to coordinate their internal functions with the clocks of the outside world, to know when to hibernate, or fly south, or sprout, or shed, or grow a winter coat. Hearts need to know when to pump, and lungs when to breathe. Different organs in the same body may keep different times to different kinds of clocks, releasing chemicals in concert with communications from a central brain. Stomachs, livers, sleep centers, may all tick to different, yet coordinated, tunes. The limits of our ability to perceive time intervals even determine how we see the world: If people could sense intervals shorter than 1/24 of a second (they can't), they would see the dark gaps between the frames of a movie; if they could perceive much longer intervals of time, they would be able to "see" plants (or children) grow.

Times change, of course, especially as we grow older. What adult isn't sometimes driven to distraction by a child's frenetic pace, and what child doesn't lose patience with his parent's slow-motion, hippopotamus approach to things? Many people have surmised that our quickening sense of time depends upon the diminishing percentage of a lifetime that each hour or year takes up as we age. To a year-old baby, a year is a lifetime—eternity. To a ten-year-old, a year is but a tenth of his lifetime, and each hour is proportionately shorter. "When he reaches fifty," writes Murchie, "time is passing five times faster still, clocks have begun to whiz, and a year is but 2 percent of his life. And if he reaches a hundred, it's 1 percent. His old friends have been dying off at a fearful rate while new children sprout into adults like spring flowers. Strange buildings pop up like mushrooms. A whole year to him actually consumes less conscious time than did four days when he was one year old."

Counting off intervals of time is a mysterious and highly individual process. How does a plant know when to flower? What kind of clock does a tree use to keep track of the passing years? How do our internal biological clocks know when to tick? The answers no doubt lie in chemical reactions, along with the ability to recognize patterns of warmth and cold, light and darkness.

We have not come close to understanding even such questions as "when" time began, for it's impossible to talk about "when"—a concept of time—unless you assume that you are already working in the framework of time. In a sense, asking what was before the beginning of time is like asking what's on the other side of space. Oddly, the current popular model of the origins of the universe sees the Big Bang happening *everywhere* in space but at *one* particular time. Time has a beginning but space doesn't.

Time and Space

> The physical theory of relativity suggests... that physical space and physical time have no separate and independent existences.
>
> —Sir James Jeans

What is certain about time is that it can't be separated from space. Time and space are tightly woven together, not only in the extreme realms where the effects of relativity become important but also in the familiar landscape of everyday life. A year, for example, is a *distance:* the distance that the earth moves in its orbit around the sun. If the distance were longer or shorter (note that the space/time adjectives are interchangeable), the time would be longer or shorter, too. A day, of course, corresponds to the distance more or less around the earth's circumference—and an hour is just a fraction (1/24) of that distance. The swing of a pendulum, the vibration of a quartz crystal or atom, anything that "tells time"—all inevitably also move through space. As Lincoln Barnett points out, "All measurements of time are really measurements in space, and conversely measurements in space depend on measurements of time."

Space and time are so closely linked in our everyday language that we rarely stop to think about it. People say that Miami is "three hours away" from New York. If someone asks you how far it is to the grocery store, you are likely as not to answer in terms of time: ten minutes. The child on a car trip who is anxious to know how much time he has to wait before the next rest stop is likely to get an answer measured in miles. An adult knows that he or she cannot attend meetings in Detroit and Seattle on the same morning because they are separated by too much space (unless, of course, the meeting takes place in the virtual world of the computer).

Like the relativity of time itself, the close kinship between time and space was once considered much more natural—before it was artificially severed by the requirements of the industrial age. Noon in New York or Tokyo was when the sundial pointed at noon—when the sun was highest in the sky—a measure of relationships in space. It didn't much matter whether one town's "o'clock" happened to match another's, because how would they compare times, anyway? The trip by ship or horse or train from one place to another covers space and takes time, so how would you know that the "time" at your embarkation point matched that at your destination?

All this changed, of course, with the coming of communication at the speed of light—radio, television, telephones, and modems. Now clocking simultaneous times at widely separated places is not only possible but essential. In fact, the needs of television networks have been a major force behind synchronizing time: The six o'clock news has to come on the air at exactly six o'clock all across the country, which means that "six o'clock" has to happen at the same time all across the country. Airline schedules, transcontinental teleconferences, Internet chat rooms, anything that forces people to synchronize their watches in different places drives another wedge in the natural affinity between space and time.

Ironically, however, it is also communication at light speed that makes the connections between space and time especially dramatic. A light-year, for example, is the distance covered by light in one year, and it is the most useful measure of distances to stars. But it is therefore obvious that looking *out* into space also means looking *back* into time. When you look at a star 5 million light-years away, you are looking at 5-million-year-old light. You are seeing the star as it looked 5 million years ago. It left its source long before modern human beings walked the earth. The light is only reaching us now, but for all we know the source is long dead; the star may be dark.

When an astronomer sees a quasar (or quasi-stellar object) 10 billion light-years away, near the edge of observable space, he or she is also seeing it as it looked 10 billion years ago—near the edge of observable time. What's been happening to that quasar during the past 10 billion years is anybody's guess. Perhaps it cooled and a speck of it condensed into a solar system with an earth much like ours, where humanlike beings are looking at us the way "we" were 10 billion years ago. Of course, asking what happened "during" those 10 billion years has no real meaning. Ten billion years ago on the quasar is today here. Usually, we use the term "during" to connote a passage of time, but in this case it obviously refers to a passage of space. An event that happened 10 billion years ago on the quasar and an event that happens on earth today are strangely simultaneous.

The notion of simultaneity, however, is yet another one of the everyday absolutes that relativity did away with. Simultaneity is relative. Imagine you are sitting somewhere out in space, and along comes a large transparent room moving past you at almost the speed of light. There is a lightbulb on the ceiling in the center of the room and a person sitting on a chair underneath it. Say the light flashes. The person in the room will see the light hit all four walls of the room simultaneously. But you will see something quite different: Since the back wall of the room is traveling toward the light at almost the speed of light, it encounters the light sooner; you see the light hit the back wall first, the front wall later. So a determination of whether or not two events are simultaneous depends on how you are moving. And since we are all in one way or another always moving through space, whatever happens to us at a different time also necessarily happens to us in a different space. It's something like the strobe photographs of a moving dancer that show his motions through space as they move through time. The fourth dimension, time, is as connected to the dimensions of depth and width as the dimensions depth and width are to length.

All this brings up the interesting question, "When is now?" Clearly, asking "when" now is makes no sense unless you also define "where" now is. The now is truly the here and now. You almost always define "now" in relation to yourself, but that may not be the same "now" for someone else in another place. Even when you look at people across the room, you are seeing them as they were a few instants ago—and if they died, like a star, in that interval, their deaths wouldn't happen for you until their light got to you.

All the various ties that bind space and time are only further pieces of evidence that seemingly isolated bits of the universe are firmly attached underneath. Space and time are linked most directly by the absolute speed of light, because light is the fastest messenger in the universe. So the three concepts fit together neatly: In order to measure speed, you need to measure distance and time—which is what speed means. But to clock speed between two distant points, you have to make sure that your clocks are synchronized. The only way to do that is to send signals via light, and still you have to account for the time it takes the light to travel. So you first have to determine the speed of light. And so on.

If it all seems curiously circular, that's only because it's all very connected. The truly curious thing is that experiment after experiment has shown that the speed of light is an absolute quantity, no matter how it's measured or who is measuring it. No matter what the motion of the observer or the observed. No matter whether you're speeding toward the source of light or rushing away from it.

At the same time (or actually a century later), countless other experiments have confirmed that measures of space and time are not absolute but depend on things like motion, or position in a gravitational field. So that the *theory* of relativity is in truth grounded in *experiment.* Indeed, the *theory* was developed in the first place in part to explain experimental *facts.* Some people think that relativity is just an esoteric set of equations of interest only to physicists and mathematicians. But even though it may not always be perceivable, relativity is a fact of life.

Special Absolutism

> A windup toy unwinds and lightens by one billionth the weight of the dot of an i. . . . The sun shines for 1 second and loses the weight of 2 dozen ocean liners.
>
> —PHILIP AND PHYLIS MORRISON, *The Ring of Truth*

Actually, there are two theories of relativity: special and general. What this means is simply that Einstein first developed his theory to fit a special case, that of steady, unchanging motion like the motion of the ship in Galileo's example of relativity. This is "special" relativity. General relativity applies to Einstein's expansion of the ideas in special relativity to all kinds of motions—in particular, changing or accelerating motions like those of objects falling under the influence of gravity.

General relativity is all about gravity and curved space and black holes. Special relativity is all about time dilation and $E = mc^2$. Both special and general relativity are theories of absolutism, because both are based on things that *do not change* in nature rather than on things that do. When you say that something is relative, you usually mean that the way it looks depends on your point of view. The point of the two relativities is that the fundamental truths of nature look the same from *every* point of view.

The question people most commonly ask about special relativity is undoubtedly, "Can I really get younger if I travel at the speed of light?" The answer is no, you can't get younger. But you can age more slowly than a friend (or twin) traveling at a slower speed. On the other hand, you pay a price: You also might temporarily get more massive in the process. The relative effects that flow from the absolute speed of light and other natural laws are as follows:

First, as you travel close to light speed, time slows and space contracts. You can imagine why this would have to be true. Suppose you are traveling toward a light source (star, candle, flashlight) a million miles away at close to the speed of light—say, 186,000 miles per second. At the same time, the light from the source is traveling toward you at 186,000 miles per second. If you measure the *same* light speed whether you are speeding toward the light or standing still (and you do), then obviously something strange has to happen to the space between you and the light source, or else to the time between you and the light source, or both. And it does. Even a clock sent around the world in a commercial jetliner (hardly traveling at light speed) comes home running a little bit slow compared to a twin clock left home "stationary" on the earth.

Second, objects traveling at close to light speed become more massive. This is not as strange as it seems. In fact, all it means is that one kind of energy is being converted into another. Long before Einstein, people knew that the energy "potential" (or potential energy) contained in an object held high off the ground could be converted into other kinds of energy. The potential energy in a pendulum bob is converted into energy of motion as it swings, then back into potential energy at the high point of the swing, then back into motion energy, and so forth. The potential energy of an apple hanging from a tree is converted first into motion (or kinetic) energy as it falls, and then into heat energy as it hits the ground and stirs up the dirt molecules. Rubbing two sticks together to start a fire is one way of converting mechanical energy into heat energy. Einstein's quantum leap of the imagination was his realization that matter itself was a form of energy that could be converted into other forms of energy in quantitative ways. Matter, if you will, is a kind of "frozen" energy.

The conversion of matter into energy is an everyday phenomenon: Every time you light a fire or burn coal, you are turning the energy of matter into the energy of heat. If you were to weigh all the molecules in the wood and the air that make the fire *before* you burn the wood and then *after* you burn the wood, you would find that the ingredients get lighter in the process of burning. But the missing matter doesn't merely go up in smoke. It is transformed into a precise amount of energy, calculated by the equation $E = mc^2$, where E is energy, m is mass, and c is the speed of light. The speed of light (c) squared is quite a formidable number, which explains why you get so much energy at the expense of so little mass. Nuclear bombs release a lot more energy for the small amount of mass they consume, because the energy bound inside the nucleus is much more "energetic" than the chemical energy of fire. The sun is also fueled by nuclear reactions. It radiates trillions of tons of its mass into space every day in the form of light energy.

The conversion of energy into mass is not as familiar, but just as frequent. In fact, every time you run, you put on a little extra "weight," or mass. And a tightly coiled spring weighs more than the same spring relaxed, due to the weight of the energy that coiling puts into it. In giant particle accelerators, electrons pushed to 99.999 percent the speed of light gain forty thousand times their original mass—which means that the accelerators are really misnamed: They are not so much in the business of *accelerating* the particles to high speeds as they are in the business of *building them up* to more formidable masses.

This helps to explain the always puzzling (at least for me) idea that some "particles" have no mass whatever. Massless particles tend to travel at the speed of light, and so contain all of their mass in the form of motion energy. Still, the motion energy of a light particle, or photon, is matter enough that it will fall under the influence of gravity just like any other object. A massless photon—like a massive bowling ball—would be pulled toward a flat earth in a smooth parabolic curve. The fall of a light beam is hard to detect only because the photon also travels *186,000 miles* horizontally for every second it falls.

Finally, the energy/matter connection is the reason behind the observable fact that light speed is the speed limit of the universe. No energy or information can travel faster than light, because as anything begins to approach the speed of light it gains an ever increasing amount of mass. Mass is a measure of inertia, the resistance to a change in motion. So the more speed or motion something has, the harder it is to make it go faster, because it also has become more massive. Eventually, the thing gets infinitely massive, which means that it would take an infinite force to make it go any faster. So even inertia is not absolute: Inertia increases the faster you go.

The reason that relativity got confused with relativism in the first place is that so many of the effects that flow from it *are* relative. They look very different from different points of view. If some people are moving past you at close to the speed of light, they will seem to get very massive. However, they will not feel themselves getting massive. They will see *you* getting massive.

Stars, planets, objects, people—anything that moves quickly relative to you appears to get more massive. The increase in mass is based on motion: The greater the speed, the greater the increase in mass. But motion, as even Galileo observed, is relative. A judgment about who is moving and who is not depends on your frame of reference, your point of view. So a judgment about how massive something is must also depend on your point of view. What holds for mass must also hold for energy, of course. If the energy of motion at high speeds can be converted into matter (and vice versa), then obviously how much energy you have depends on how much motion there is. So energy in this sense is relative, too.

The same lopsided view of things also applies to time and space. As we have seen, travel at high speeds causes your clocks to tick more slowly and your space to contract. But these effects are necessarily relative to something else. If your friend stays home while you whiz off on a quick trip around the stars, you may return to earth still "young" years after your friend has died at a ripe old age. But according to your own internal biological clocks, or the watch on your wrist, or any other kind of timekeeping device on your starship, time flows as always. You can't tell that "time has slowed down," because even your brain is running slow, your heart and lungs are running slow. Even radioactive clocks would run slow. All clocks run relativistically, because relativity is a property of time and not of clocks. (Or perhaps I should say, *as well as* of clocks.)

The result is that you have no way of knowing whether you are more massive or not, whether your time has slowed or not, whether your space has contracted or not, even whether you are moving or not—just like a person on Galileo's ship. Your mass is normal mass, your time normal time, your space normal space. Einstein called the view that you yourself see the "proper" view, to distinguish it from other frames of reference. Everything in your view seems appropriate and proper, while things and people moving relative to you appear slow-moving, squashed, and somewhat twisted. This is certainly a familiar phenomenon in everyday life: We tend to view our own habits and customs as normal; only the behavior patterns of other people seem "warped."

Despite all this superficial relativism, however, *what actually happens* remains remarkably absolute. Things look different from different points of view, just as a box looks different depending on whether you are standing still next to it or speeding by it in a passing car. But the box itself does not change. And the laws of nature do not change. The relationships between things and events do not change. That is why you cannot tell whether or not you are moving, whether or not your clocks are slow.

One of my favorite examples of this absolutism underlying relativity comes, rather unexpectedly, from the realm of biology.

"Usually, we pity the pet mouse or gerbil that lived its full span of a year or two at the most," writes Stephen Jay Gould.

> How brief its life, while we endure for the better part of a century. [But] such pity is misplaced.... Their lifetimes are scaled to their life's pace, and all endure approximately the same amount of biological time. Small mammals tick fast, burn rapidly, and live for a short time; large mammals live long at a stately pace.... All mammals, regardless of their size, tend to breathe about 200 million times during their lives (their hearts... beat about 800 million times).... Measured by the internal clocks of their own hearts or the rhythm of their own breathing, all mammals live the same time.[11]

From the confines of our own limited perspectives, things may seem very different where in truth they are very much the same. The equations of relativity provide a kind of language—or better, a dictionary—that translates from one frame of reference into another.

General Absolutism

> The gravitational field exists for the outside observer; it does not for the inside observer.
>
> —Einstein and Infeld, *The Evolution of Physics*

General relativity extends this idea from steady motions—light beams silently passing each other in the night—to changing or accelerating motions. (Physicists use the term *accelerate* to mean any kind of change in motion—not just going faster but also going slower, stopping, or changing direction.) The most famil-

---

[11] There are exceptions, of course—notably people, who due to a strange kind of perpetual immaturity live longer than other mammals of their size... and are living longer and longer.

iar accelerating motions in the universe are those connected with the force of gravity. Falling objects fall faster the farther they fall, and planets orbiting the sun accelerate in the sense that they are constantly changing direction. The two relativities were part of Einstein's lifelong attempt to unify all of nature's forces: Special relativity rested on the relative motions of electric and magnetic forces, or light. General relativity brought the final then-known force—gravity—into the family fold. The basic approach behind both relativities was much the same.

Like special relativity, general relativity follows from what *you cannot tell,* from what *does not* make a difference. In special relativity, you cannot tell whether you are moving steadily or at rest. In general relativity, you cannot tell whether you are accelerating or standing in a gravitational field. The two situations are exactly equivalent, something Einstein understandably called the equivalence principle.

Imagine (as Einstein did) that you are in an elevator and the cable snaps: Suddenly you are in a situation with no gravity. Drop a ball and it floats in front of you; put out your arms and they float at your sides. You are in a situation of free fall, no different in any way from the zero-gravity environment of outer space. As long as you are moving along with the acceleration of gravity, the force seems to disappear, just as the force of magnetism seems to disappear when you "travel along with" an electron. In this free-fall situation, is there any way you could tell whether you were in zero gravity or moving along with the acceleration due to gravity? No.

Now imagine you are floating around in a rocket ship in space, and suddenly the rocket ship begins to go faster and faster. Drop a ball and the floor of the rocket ship will quickly catch up with it. The ball does not "float" but "falls." Can you perform any kind of experiment that will tell you whether you are really in a rocket ship accelerating in space or just sitting in the same rocket ship in a launching pad on earth under the influence of gravity? No.

General relativity solved (indeed, it was inspired by) the great riddle about the relationship between inertia and mass: Why do bowling balls and Ping-Pong balls fall at the same rate in a vacuum? Because if they were dropped in an accelerating rocket ship, the floor would catch up with them at the same time. Either way, the two would "hit the ground" simultaneously. If you were living in a gigantic, earth-size accelerating rocket ship, you might think that things around you were held to the ground by the "force" of gravity. But an outside observer might see that the reason things were held to the ground was really that the ground was accelerating toward them. According to general relativity, the force of gravity is relative.

When Einstein first realized this, he was understandably jubilant. He wrote:

> At that point there came to me the happiest thought of my life, in the following form: Just as is the case with the electric field produced by electromagnetic induction, the gravitational field has similarly only a relative existence. *For if one considers an observer in free fall, e.g., from the roof of a house, there exists for him during his fall no gravitational field*—at least in his immediate vicinity. For if the observer releases any objects they will remain relative to him in a state of rest, or in a state of uniform motion, independent of their particular chemical and physical nature. (In this consideration one must naturally neglect air resistance.) The observer therefore is justified to consider his state as one of "rest."

Many aspects of general relativity are hard to get used to, and not only for laypeople. As Victor Weisskopf explained his feelings on the subject: "It's like the peasant who asks the engineer how the steam engine works. The engineer explains to the peasant exactly where the steam goes and how it moves through the engine and so on. And then the peasant says, 'Yes, I understand all that, but where is the horse?' That's how I feel about general relativity. I know all the details, I understand where the steam goes, but I'm still not sure I know where the horse is."

Equivalence itself is not that hard to get used to. When astronauts accelerate toward space in their shuttles or moon rockets, they measure the force of acceleration in so many G's—G being the designation for one earth's gravity. Two G's equals twice the force of earth's gravity, and so on. Visions of permanent space stations in the sky always substitute another kind of acceleration for gravity—centrifugal force. Huge wheels spin in space, throwing people, houses, and everything else outward like stones twirled on strings. If the "ground" is built on the inside outer wall of the wheels, then the centrifugal acceleration will be exactly equivalent to gravity.

Curiously, Newton used the example of centrifugal acceleration to prove that accelerating motion was *absolute*, not relative. He said that while steady motion could obviously be relative (à la Galileo), accelerating motions were entirely different. If you spun a bucket of water, the centrifugal force would make the water rise at the sides, in the same way that the spin of the earth causes it to bulge outward at the equator. The bulge was clear evidence that these things were moving and not at rest. But around 1900, Ernst Mach pointed out that if you spun the whole universe and kept the bucket or the earth at rest, then the result would be the same, so you *still* could not tell whether you were at rest or accelerating.

It's certainly odd to imagine that a force such as gravity can be relative. When you push something to make it go—throw a ball, for example—there doesn't seem to be anything relative

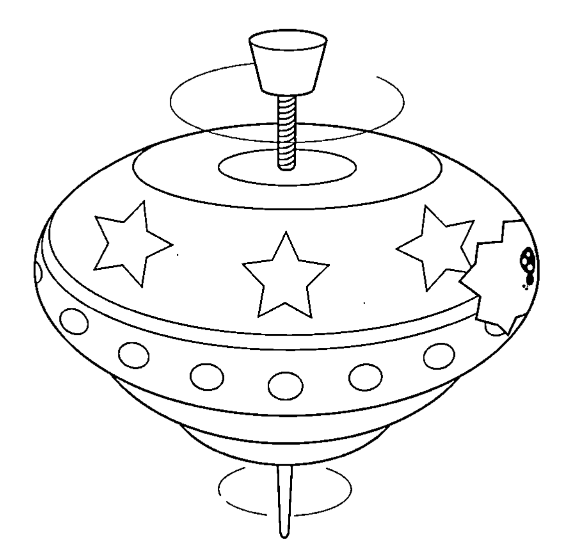

The ant inside the wheel feels a force like gravity pulling it "down"—which is outside; "up" to the ant is toward the center of the wheel.

about it. But consider this: One of the measures of a force is the motion it produces. Push something with a large force and it goes farther (and perhaps faster) than it goes when you push it with a small force. If motions are relative, however, and motions are the result of forces, then it is not so surprising to find that forces have relative qualities, too.

Rather than a force, general relativity describes gravity as the curvature of space-time itself—potholes in the unseen pavement of the universe created by the weight of heavy objects. Stars, planets, and even intense concentrations of energy ($E = mc^2$) all warp space and time. Objects passing nearby these dips in space-time feel an irresistible attraction, frequently "falling" in, like the wheel of a car sinking into a hole in the street.

Under extreme circumstances, the hole may be a one-way street for everything that approaches. A black hole, for example, is created when space bends back on itself so severely that nothing, not even light, can escape. Black holes—as well as wormholes and other proposed exotica—follow directly from Einstein's conclusion that gravity is relative; it isn't a force that "forces" things to fall but rather a property of space. Things fall because that is their normal "straight-line" path in a curved, four-dimensional space-time continuum. And curved space is much more than just another way to look at the force of gravity. It actually gives different results, which have been verified experimentally. While curved space is invisible to human eyes on human scales, its distorting effects on starlight from distant stars and galaxies are crystal clear. In some cases, large concentrations of matter create a gravitational "lens" that multiplies images of galaxies like a cosmic kaleidoscope.

For physicists like Kip Thorne who spend their careers exploring the consequences of "Einstein's outrageous legacy," as Thorne calls it, curved space isn't any harder to imagine than the stars or particles that other physicists study. After all, he points out, "you can't see atoms with your eyes [either]. You can't see air with your eyes."

For most of the rest of us, curved space remains deliciously paradoxical. What, after all, is "straight" space? A straight line is the shortest distance between two points, but what is that? Usually, by "shortest distance" we mean line of sight, or the path of a light beam. But we *assume* that light travels in straight lines. If the light beam curves, is it going straight in curved space? Or is it curving in straight space? There are even some cases in which the shortest distance between two points is the path that takes the longest time according to your watch—because time *slows* as you go faster!

Relativity, in short, does not mean that everything is relative. It means that appearances are relative—and you already knew that. It is not so surprising that perceptions change as you look at things from different points of view. It happens all the time. What is strange is that you can manage to reach the same conclusions from so many different points of view. Once you have discovered what is absolute, then you can learn what is only a matter of appearance.

PART III:

Threads and Knots

Those beautiful spiral patterns that one sees in pictures of galaxies are not, in most cases, the patterns of where the stars are.... The spiral patterns that flow through the disk of a galaxy are only the most visible manifestations of the workings of a continually renewing force.

—LEE SMOLIN, *The Life of the Cosmos*

CHAPTER NINE

Waves and Splashes

> So what is this mind, what are these atoms with consciousness? Last week's potatoes! That is how we can *remember* what was going on in my mind a year ago—a mind which has long ago been replaced.
>
> That is what it means when one discovers how long it takes for the atoms of the brain to be replaced by other atoms, to note that the thing which I call my individuality is only a pattern or dance. The atoms come into my brain, dance a dance, then go out: always new atoms but always doing the same dance, remembering what the dance was yesterday.
>
> —RICHARD FEYNMAN

FEYNMAN'S WORDS express what may well be the oddest bit of enlightenment to pass down to us from the hard world of physics: The "dance" is more real than the atoms. The "abstract" patterns of the physical universe are more concrete than the things you can feel or touch. What makes matter and force solid and permanent are mostly intangible repeating rhythms sung by an ever changing chorus. Atoms come and go, but memories can persist for a lifetime. The law of gravity sucks stars and planets into spheres regardless of their ingredients. Quantum mechanics makes gold gold, wherever it is, and wherever it's been.

Physicists look at patterns for clues to underlying forces—as do parents, psychologists, economists, and occasional politicians. It's a common mistake to dismiss the pattern as something of no real substance, while seizing on seemingly concrete individual cases. Patterns are more real than rocks or atoms or black holes. Patterns are us.

According to conventional scientific wisdom, every atom in the human body gets replaced within a seven-year period. If your "self" is the atoms that make you up, then you become someone else every seven years. You're not even yourself from one day to the next, or even from moment to moment.

Clearly, we don't define people by their atoms. "If O. J. [Simpson] had tried to get off by mounting appeals for seven years, then saying, 'Gee, you've taken so long to convict me, I'm not the same person,' that would be [absurd]," says Stanford philosopher John Perry, who specializes in the study of self. "We're not trying the atoms, we're trying the person."

Or take something even more tangible. Take a chair, says Feynman:

> Philosophers are always saying, "Well, just take a chair, for example." The moment they say that, you know that they do not know what they are talking about any more. What *is* a chair. Well, a chair is a certain thing over there... certain?, how certain? The atoms are evaporating from it from time to time—not many atoms, but a few—dirt falls on it and gets dissolved in the paint; so to define a chair precisely, to say exactly which atoms are chair, and which atoms are air, or which atoms are dirt, or which atoms are paint that belongs to the chair is impossible.

Eddies and whirlpools and raindrops are patterns of water molecules (it does not matter *which* water molecules) that exist almost independently of the water itself. The water flows *through* the forms.[12] A rainbow is not a thing but a pattern of light re-

[12]A beautiful book on this subject is *Patterns in Nature* by Peter S. Stevens.

fracted from water droplets—*different* light from *different* droplets for every person who sees it. Each person sees a rainbow unique to a personal perspective; only the consistent pattern of colors in a curved bow deceives people into thinking it's a concrete "thing." The water molecules themselves have properties that depend on the pattern of oxygen and hydrogen atoms that make them up. All the qualities that make water the fountain of life—the stuff of blood, sweat, and tears—come from the arrangement of these atoms, the same arrangement that is reflected in the shape of every snowflake and soap bubble.

Even a galaxy of stars is largely an abstraction, in the sense that the individual stars in the spiral arms are continually replaced by new ones. The pattern of stars that makes the spiral even rotates at a different speed than the stars themselves. The sun currently resides between spiral arms, but once it was spinning out on a limb, and (if it lives long enough) may well someday migrate to an inner limb, a home for middle-aged stars.

Abstractions seem magical because they can exist independent of matter—and also because they can do things that matter itself cannot do. Family traits and traditions can long outlive any individual family member, just as a comet's tail can sweep around the sun faster than the speed of light. It can do this because, as Guy Murchie points out, "a comet's tail does not remain the same tail any more than a stream of water from a hose remains the same water."

The laws of nature are really observed patterns—relationships between things and events. They have great power because they allow for infinite variety amid amazing regularity. All people—like all planets and all trees—are cut from the same pattern, yet they exhibit a wide range of individual forms. We are all alike, yet all different. These patterns or "laws" are often expressed as mathematical formulas—something that allows them to reach out and explore far beyond the realms of human experience. A pattern, unlike a person, can safely go to extremes. It can tell you what will happen if you leave your money in the bank for a million years, or what life on earth might have been like a million years ago. It can both extrapolate and interpolate. It can help you to find out where you're going and where you've been. It can tell you what takes place inside an atomic collision and what matter is like at infinite gravity, as in a black hole.

Perceiving these patterns is so important to human survival that we often see patterns that "aren't there"—in clouds, in cracks in the ceiling, the "man" in the moon. Actually, these patterns are highly subjective. In China, it is a "rabbit" in the moon. The configuration of stars we call the Big Dipper is interpreted as the Plough in England. But one way or another, almost everyone is involved in looking for patterns that add up to a larger reality. Doctors look for patterns of symptoms that spell disease, journalists look for patterns they call social trends, entrepreneurs look for patterns in the marketplace, and scientists look for patterns known as laws of nature.

One particularly persistent pattern in nature is the event, and the object, we call a wave. What is a wave? A wave is necessarily a kind of vibration, but you can have a vibration that doesn't make waves. (The vibration of a bell in space doesn't make waves, because there is no air in space to carry the sound.) Normally, however, any kind of twang or disturbance in the nature of things spreads out in far-reaching waves of influence. A splash, in contrast, is a one-shot affair. The Edsel made a splash; Elvis and Einstein made waves.

The difference between a wave and a splash is that a wave is a lot larger than itself. It can separate itself from the original disturbance that created it and carry information far from its source, bending around corners, going right through things, sometimes capsizing people or even whole countries in the process.

Once out on its own, its strength does not depend on any particular event but has a power that sometimes exceeds all expectations. It can interact with other waves in ways that make it grow to monstrous proportions—or completely disappear.

A wave can do all these things because it is not made of "stuff"; it is a movement of information. A fashion wave can start with a splash, for example. But once set in motion, the wave moves quite independently of the people caught up in it. The people are only the carriers. The wave itself consists of a pattern of how people pick up the fad and then drop it. The people stay where they are; only the wave spreads. In the same way, waves of thought and feeling spread throughout your body using nerves as electrical conduits. But the nerves themselves stay put. Most waves, in fact, die out rather quickly. Only when there is a continual input of new energy—people picking up a fashion wave, wind fanning ocean waves, fresh sources of electrical energy nudging along nerve signals like so many telephone cable repeaters—do they keep going or even gain strength.

Some splashes make more than one wave. When you drop a stone in water, the air next to the water receives a push, which it passes along to the next bit of air, and so on and so forth like a pulse sent along a Slinky or the wave of motion (and emotion!) that one car nudging another can send through a traffic jam. The push of air arrives at your ears and is heard as sound. At the same time, the splash starts a wiggle in the water. The up-and-down motion travels through the water, incidentally carrying up and down any sticks, leaves, ducks, or boats that happen to be in its path. The sticks and boats won't travel to the opposite shore any more than do the particles of water. The wave moves *through* the water, like a rumor through a crowd.

What makes waves, in other words, is not traveling stuff but moving signals. Light waves and sound waves carry voices, words, and images. Ocean waves carry information about storms far out to sea, and a tidal wave brings the message that somewhere the earth has moved. The first domino in a row transmits the fact that it has fallen to the last domino without ever moving itself. What has propagated through the row is a change in the domino's *position*—from vertical to horizontal. What has moved is a *condition,* a state of affairs.

That's one reason waves can interfere with each other and still hang onto their own identity. Two sets of waves interfere constructively when crest meets crest and trough meets trough, so that the effects of the waves add up. When two sets of waves interfere *destructively,* however, one is moving up while the other is moving down; crest meets trough and the two motions cancel each other. The result is nothing.

Whenever two somethings add up to nothing, it is a sure sign that you are dealing with waves. Two houses, or two pebbles, cannot add up to no houses or no pebbles. The fact that two light beams interfere with each other in ways that produce bands of darkness stood as firm evidence for centuries that light had to be a wave—until Einstein came along and once again kissed to life Newton's conviction that light also had definite particle properties. Twenty years later, people discovered that all particles—electrons, protons, neutrons, and so on—also exhibit interference effects. So if there is something about a light wave that acts like a particle, there is also something about particles that acts very much like waves.

As long as two such patterns repeat themselves with the same frequency, they will always interfere either constructively or destructively. Like two people marching at the rate of precisely ten steps per minute, if they start out in step they will stay in step, and if they start *out* of step they will stay out of step.

This is not, in general, the way things behave, however. Normally, two closely related patterns tend to get in and out of step, so that in some places they will interfere constructively and in other places destructively. Periods of loudness alternate with periods of silence; areas of brightness alternate with areas of darkness. The result is the audible beats that tell when two instruments are slightly out of tune—the colorful bands in soap bubbles and oil slicks, opals and butterfly wings.

Interference is a larger pattern that emerges from the superposition of two other patterns. It therefore makes a very useful magnifier. Interfering laser light is routinely used for surveying landscapes (even for measuring the distance to the moon), and the interference of X-ray light is used to study the nature of crystals. The interference patterns of radio signals from quasars billions of light-years away have been employed to measure the minute motions of continental drift. Most new telescopes (and even older ones) are being outfitted with interferometers—mechanisms that combine the light waves from two different "receivers" to produce a much sharper, grander image.

Some physicists even describe the quantum states of the atom as a kind of interference pattern of particle waves. That is, where the particles (wavicles?) interfere constructively, it amounts to a stable state. Since all particles in nature are associated with wave patterns, all the atoms and molecules composed of them are patterns of such patterns. Perhaps physicist J. Robert Oppenheimer was thinking something along these lines when he said that the interference of electron waves gives rise to many "novel effects.... It is responsible for the permanent magnetism of magnets. It is responsible for the bonding of organic chemistry and for the very existence in any form that we can readily imagine of living matter and of life itself."

The fact that an electron is a wave, however, does not preclude its also being a pointlike particle—which all experimental evidence shows it to be. Matter waves are not *distribution* waves—that is, waves that are distributed over space. Rather, they are *probability* waves, which chart the probability of a particle's being in a certain place at a certain time. Like a wave moving through a row of dominoes or even a field of wheat, what is distributed in such a wave is not matter but a state, or condition—a wave of information. This fits in with the forty-year-old observation of Sir James Jeans that electron waves are essentially "waves of knowledge." The wave charts the probable place where a particle would appear *should we choose to measure it.*

When they're not interfering in each other's affairs (and even after they do), most waves move right through each other like so many ghosts. Sound waves, light waves, bow waves from boats, are continually crossing and interfering, yet surprisingly enough they arrive at their destinations essentially unchanged and intact; if they didn't, visual and auditory images would be tangled in an impenetrable web of white noise.

Not all waves are started by splashes, of course. Sand dunes and snowdrifts and even the waves that flow through flags and wheat fields are shaped by the force of wind. Light waves are shaped by the ebbing of an electric field that creates a flowing magnetic field that, when it ebbs, creates another electric field, and so on. Many ocean waves are propelled by the same pull of gravity from the moon which creates the tides. The shape of waves, like the shape of planets and soap bubbles, the twirl of a tornado, the six-fingered star of the snowflake—all are sculpted by relatively few forces and motions. And if some patterns seem to repeat themselves, it is because the forces that form them are strong currents that flow throughout nature.

Indeed, a particular kind of wave tends to pop up everywhere you look—the sine wave. A sine wave is the visible manifestation of a very fundamental kind of motion—the motion of a pendulum. But it's also the basis of a rabbit-hole system of transportation devised by Lewis Carroll while he was dreaming up *Alice in Wonderland.* It works like this:

If you fell down a rabbit hole that went clear through the earth, you would be pulled toward the center by gravity until you reached a maximum speed of about five miles per second. Once you passed the midpoint, gravity would start pulling you backward (because most of the earth's mass would be behind you), but your own momentum would keep you going just about until you reached the opposite side—say, Australia. If you forgot to climb out at Sydney, you would be pulled back again toward your starting point, and you could keep swinging back and forth like a human pendulum until friction slowed you down.

The beauty of the system is that any rabbit-hole trip through the earth would take the same length of time—exactly forty-two minutes. Whether you jumped into an eight-thousand-mile rabbit hole to Sydney or a four-thousand-mile rabbit hole to Prague or a two-thousand-mile rabbit hole to Miami Beach, you would still arrive at your destination exactly forty-two minutes later. You would never get going quite as fast during the Miami trip, because the acceleration due to gravity wouldn't be as great; but then again, you wouldn't have to go as far.

This explains why a pendulum (within limits) always takes the same period of time to swing back and forth even as the swings get smaller. The force that pulls the pendulum toward the center of the swing increases as the distance from the center increases; so as the distance gets larger, so does the force, and everything evens out in the end.

A sine wave is a picture of this motion. And when a stone drops in water, it creates a wiggle with essentially these same properties—sending out water waves and sound waves (not to mention light waves), all in the shape of sine waves.

Waves are just one of the common patterns that result from the relatively few forces and motions that mold nature. Gravity is particularly powerful in this respect. Once Newton realized that it shaped the planetary orbits, "a lot of other things became clear," writes Feynman. "How the earth is round because everything gets pulled in, and how it is not round because it is spinning and the outside gets thrown out a little bit, and it balances; how the sun and the moon are round, and so on." Stars and planets are round because gravity pulls matter toward other matter—a direction people on earth parochially call "down."

The parabolic path of falling water is shaped by the pull of gravity.

The very shape of curved space is simply the pattern of the way things "fall" under the influence of a gravitational field, in the same way that iron filings fall into a certain shape when they come under the influence of a magnet. The formulas that define these forces are in one sense mathematical expressions of *behavior* patterns. Chairs and black holes, atoms and ants, honeycombs and human bones, all take shapes that fit the pull of nature's forces. In his famous book *On Growth and Form,* D'Arcy Thompson said that any object was essentially a "diagram of forces."

Patterns may seem ephemeral, but in the end they are the enduring essence of things. They are the waves of substance that linger long after the momentary splashes of fate and fashion have gone silent.

CHAPTER TEN

Sympathetic Vibrations

> He put the little vibrator in his coat pocket and went out to hunt a half-built steel building. Finding one in the Wall Street district, ten stories high, with nothing up but the steel-work, he clamped the vibrator to one of the beams. "In a few minutes," he told the reporter," "I could feel the beam trembling. Gradually, the trembling increased in intensity and extended throughout the whole great mass of steel. Finally, the structure began to creak and weave, and the steelworkers came to the ground panic-stricken, believing that there had been an earthquake. Rumors spread that the building was about to fall, and the police reserves were called out. Before anything serious happened, I took off the vibrator, put it in my pocket, and went away. But if I had kept on ten minutes more, I could have laid that building flat in the street. And, with the same vibrator, I could drop the Brooklyn Bridge in less than an hour."
>
> —MARGARET CHENEY (on Nikola Tesla), *Tesla: Man out of Time*

PERHAPS NO SUBJECT from the house of science has so infused the language of everyday life as the notion of resonance, a persisting pattern of harmonizing waves. People speak of being in tune with the times or out of tune with each other. They speak of sympathetic vibrations, and of being on the same wavelength. They speak of ideas that resonate and anecdotes that ring true.

In the physical world, resonance is equally pervasive. From

the rings of Saturn to the colors of rainbows to the brief lives of subatomic particles, resonance rules. Indeed, physicists who search for the fundamental building blocks of nature routinely refer to particles as "resonances," and sometimes describe their work as "resonance hunting."

Literally, resonance means to resound, or sound again: to echo. It is the synchrony of many small periodic pushes that work in unison to add up to a much larger one. Putty doesn't resonate, because it is too full of internal friction to vibrate; a dropped handkerchief simply falls. In order for something to resonate, it needs a force to pull it back to its starting place and enough energy to keep it going. The trick is to have it resound again and again, but this requires putting energy in faster than friction can take it out.

Two can play this game much better than one, because one can feed energy to the other. That's what sympathetic vibrations are all about. When people talk about resonance, they are usually referring to the confluence of more than one action: the escapement of a clock that lets the spring go just enough and at just the right time to give the pendulum (or crystal) the push it needs to keep it going. Or two partners in a business feeding each other ideas and energy at just the right times and places to get big results.

Putty and handkerchiefs aside, the universe as a whole is a remarkably springy place. Planets and atoms and almost everything in between vibrate at one or more natural frequencies. When something else nudges them periodically at one of those frequencies, resonance results. Soldiers marching in step with the natural frequency of a bridge can cause it to collapse, which is why soldiers break step when crossing bridges. Some studies suggest that giant icebergs are splintered by the resonant force of gently lapping ocean waves. Resonance is even responsible for the forty-foot tides in the Bay of Fundy. My friend the physicist once insisted that through a similar series of well-timed pushes on water, any determined child (or adult) could empty a full bathtub in a single cycle, once the "tide" had been built up high enough. Try this at home.

Nikola Tesla, the eccentric inventor of alternating current, became so obsessed with the power of electrical resonance that he bragged that he could use it to split the world.

The power of resonance comes from literally being in the right place at the right time. For it to work, there has to be a harmony between what you're doing and the way something (or someone) wants to go. The almost eerie purity of laser light results from the fact that all the atoms in the excited gas are poised just so that a gentle nudge of energy will cause them to give off light in patterns exactly aligned with each other. In much the same way, the gravitational pulls of Saturn's four inner moons happen to be in harmony with the natural period of rotation of some particles in the rings at certain distances from the center. This period at one point coincides exactly with one-third of one moon's period, one-half of another moon's period, one-quarter of still another's, and so on. The combined force is enough to push (or pull) all particles out of those places, creating the gaps in the rings. In fact, if it weren't for resonance, they wouldn't be rings at all, but a single solid disk.

Resonance, in other words, allows a lot of little pushes in the right place to add up to big results. Particle accelerators use this principle of "a well-timed kick in the pants," as one physicist put it, to nudge electrons or protons almost to the speed of light. It is also a familiar phenomenon in everyday life. A lot of little pushes in an already angry crowd can lead to a full-scale riot. A lot of little digs over the dinner table can lead to divorce.

But resonance is far more than a brutal amplifier. It is also music to our ears. A violin bow slips along the string, catching it imperceptibly at precise intervals that push it at the proper time to keep it vibrating. The body of the instrument vibrates in tune to a rich range of harmonics. To play a flute, you set the air inside resonating at many different frequencies, depending on how far the sound waves travel between the mouthpiece and the finger holes. The holes are placed to pick out those tones that correspond to the standard musical scale, but how you blow determines whether your flute will resonate with the notes that are purest and loveliest.

One of the most useful properties of resonance is its ability to act as a precision tool, plucking one pure vibration from a sea of others, a single tune out of a confusion of noise. Imagine you walked along a path strewn with pebbles and bells, and kicked them out of your way at random. The bells would ring while the pebbles wouldn't. Why? Because the pebbles would reflect the energy of your kick every which way as they flew off in different directions. But the bells would be able to feed the energy back into themselves by virtue of their natural springiness. The bells would be able to contain the energy long enough to "ring."

These ringing bells are akin to the natural harmonics that give a unique character to the music of a violin or flute, the voices of men and women, the clatter of a tennis racket and the *plop* of the ball. When disturbed, each object resounds and vibrates only to its natural set of frequencies, which together form its special sound. All other vibrations are canceled, or sent flying off in random directions. There is no mistaking the voice of your child, because the rush of air from the lungs that starts out as just so much noise is selectively amplified by a particular configuration of mouth, nose, chest, and throat to sound in a particular way—just as the tuner on your radio amplifies only one narrow range of frequencies at the expense of all others. The others are simply scattered.

This very same property colors everything you see. Sodium lights are yellow because sodium atoms vibrate with those frequencies your brain perceives as yellow. Mercury atoms vibrate with a bluish light; neon atoms send out vibrations that reach your brain as "red." The colors are not the result of a single "note" but rather of the atom's characteristic "chord," composed of all the atom's resonant states. When atoms absorb rather than emit light, they leave behind shadows, but the notes in the "chord" are the same. White light from a star that passes through surface gases containing an element that strongly absorbs green will arrive on earth with a sharp dark line in the green part of the star's spectrum. This kind of long-distance chemical analysis has revealed that we and the stars are fashioned of the same stuff.

Down here on earth, resonant absorption colors everything from sports cars to fruit. The pigment molecules in the skin of a McIntosh apple absorb the parts of sunlight that vibrate harmonically in the frequencies we see as blue and green; the rest of the light is reflected, and we see red. Chlorophyll molecules in green leaves vibrate to the tune of red and blue and absorb them, reflecting the leftover green; the same leaves absorb green and reflect autumn colors in the fall. Ultraviolet light vibrates harmonically with the molecules in glass; the visible light gets through, but you can't get a suntan unless you open the window. The ozone layer in the atmosphere, like sunscreen, also absorbs much of the resonant ultraviolet vibrations from the sun and protects us from potentially damaging light.

Even rainbows are resonances. The colors of white light passing through a prism or a raindrop spread out because the colors near the violet end of the spectrum are more nearly resonant with the glass or water molecules than the colors at the red end. The closer a color is to perfect resonance with the glass molecules—that is, the more in tune the two vibrations are—the longer it lasts. The purer the resonance, the longer—like a good bell—it rings. Violet light "rings" longest, and therefore bends most, as it passes through a prism.

Resonance, in other words, determines what sinks in and what goes through. It makes the difference between visible and invisible, between transparent and opaque. Metals are opaque because their many freely moving electrons can vibrate to just about any frequency—and so absorb them. The ability of these same free electrons to reradiate all those frequencies explains why metals make good mirrors. On the other hand, almost everything is transparent to a radio wave, because almost nothing resonates in radio frequencies. You can hear your radio (and TV) signals right through the thickest walls.

Sometimes resonance turns things into one-way doors, or radiation traps. Glass can be just such a one-way window for light. Visible light from the sun passes through a glass window and is partially absorbed—say, by someone inside wearing a red dress. The red light reflected from the dress can pass right back through the window, but the other colors are absorbed by the dress and eventually are reradiated as the lower-frequency light we call heat. But the heat radiation can't get back through the window. It is trapped inside. The result is the so-called greenhouse effect, which is great for warming up greenhouses but dangerous for our atmosphere. An excess of hydrocarbons from the burning of oil and coal are making the sky into a one-way window that traps the sun's heat, warming our environment to possibly dangerous levels.

Most magically, resonance can make things appear out of nowhere, like rabbits out of hats. The music broadcast by your local radio transmitter seems to spring out of thin air when you tune your receiver to a sympathetically vibrating frequency. Your smoothly running car can suddenly break out in a bad case of the shakes when the cycle of the unbalanced wheel exactly matches the natural rhythm of the springs.

Some physicists use this analogy to explain how resonances can produce particles. In the elementary universe of particle physics, every energy is associated with a frequency and vice versa. It is part of the natural complementarity of matter that it has both wave and particle characteristics. Since matter has wave properties, it has frequencies, too. Each particle wave, for that matter, has a specific frequency, and that frequency corresponds to a specific energy. Energy (according to $E = mc^2$) equals mass. So in a very fundamental sense, the way something "vibrates" seems to determine what it is. And when accelerator physicists "tune" their beams of particles so that they collide with a burst of energy that vibrates at exactly 7.5 times 10^{23} cycles per second, then *presto!* They have created a particle (or, really, pairs of particles), in much the same way as you can create a tone by blowing with precisely the right energy over the top of a Coke bottle.

All analogies break down as you descend to subatomic depths, of course. But you could also imagine the particles/resonances as the bells on the pebble-strewn path. Most kinds of collisions would result in a lot of kicked pebbles, with energy and motion redistributed in all directions. Every now and then, however, you would hit something and it would "ring" for a longer time, because it would have the special property that it could feed energy to itself. You would know that there was something special about it. You might even call it a particle. It is a rather nice thought that the universe might turn out to be essentially just such a symphony of submicroscopic chiming bells.

CHAPTER ELEVEN

Symmetries and Shadows

> Men imagine gods to be born, and to have clothes and voices and shapes like theirs.... Yea, the gods of the Ethiopians are black and flat-nosed, the gods of the Thracians are red-haired and blue-eyed.
>
> —XENOPHANES

SOMETIMES PATTERNS get misplaced. We think they are signals from the world "out there," when they are really images created in the mind's eye. Frequently it's hard to tell the difference. Patterns take detours on the road from there to here, from outer space to inner mind. Patterns of light and sound, like patterns of electrical nerve impulses, get bounced around, misdirected, rerouted—making it impossible to tell where they came from, whence they carry news.

Think how often you see a reflection and assume it is a window to the outside world. Reflections are as real in their own way as the so-called concrete objects around us. You don't know where a light beam has been before it gets to you, even when its source seems to be right in front of your eyes. You don't know that the bright red spot of light hovering outside your car window is really the reflection of a bicycle taillight behind you, so you might see it as a UFO. Reflections take light and all its colorful images and spin them around as if in a revolving door. But the revolving door is invisible; you have no way of knowing how many times it's been spun around before it gets to you. As I look out my window, I think I'm seeing the edges of a pond about a half mile away. But actually I'm seeing the reflection of sunlight on the pond—a reflection created by light that has been absorbed and reemitted many times by air molecules and then bent and unbent as it passed through the window, and then again by the corneal covering of my eye. I can't see the light from the pond without also seeing all that has happened to it between there and here.

All of the information and ideas we deal with are filtered through many layers of perceptions and prejudices—our own and those of people around us. This is also true of images borne on light. My son once observed that the difference between stars and planets was that stars are "pointy." But the broad spokes of light that fan out from stars and streetlights and Christmas tree bulbs are not produced by the stars at all. A star is a great round ball of gas. (The sun is a star.) The spokes or "points" are created as the light bends around onionlike layers of transparent cells that make up the lens in your eye. The stars don't really twinkle. They only seem to shimmer, because the air between us and the stars is constantly moving, shaking the light from the star around. The sparkle that we see when we look at the stars is a twinkle in the mind's eye.

Astronomers and particle physicists constantly have to worry about distortion and digression of the patterns of light and sound they use to decode the physical world. Is the light from that star really dim, or has some of it been absorbed by interstellar dust? Is that signal coming in a straight line, or has it been bent or stretched by some gravitational field? Did that particle track come directly from the center of the collision, or was it reflected?

A reflection is a particular type of digression that is remarkably omnipresent in nature. Almost everything reflects something from the proper angle. An echo is the image of your voice

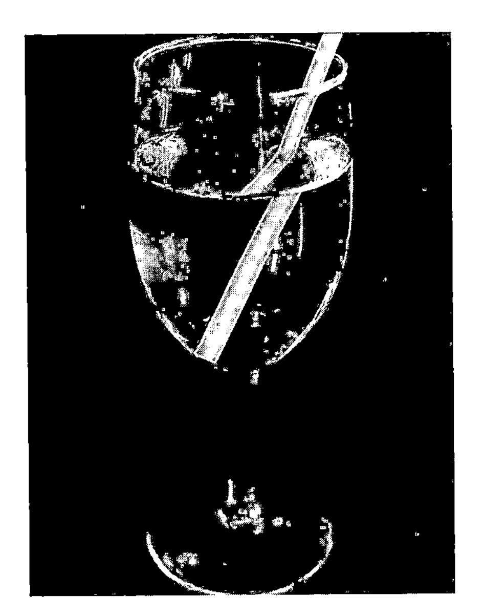

As light bends from water to air, it takes its images with it, creating all manner of "miracles."

as it reflects from a canyon wall. Water waves bounce off the shoreline, reflecting its contours back into the ocean. You can make a mirror out of a book or desktop by peering over its edges at a very shallow angle—just as children learn to make mirrors out of ponds when they skip stones. Guy Murchie tells of a lake in the Alps so smooth that marksmen aiming their bullets at the water are able to hit their targets on land.

Just about everything we see or hear, for that matter, is reflected from something else. Not all these reflections are mirror-like (like echoes). But anything that doesn't glow with its own light must bask in the light of something that does. When you turn on a lamp in a dark room, the light bounces around from the walls to the furniture and finally to your eyes. Everything you see in the room is really reflected lamplight. Even the red of the couch is a reflection of the red hidden in the spectrum of the white light from the bulb; when the white light hits the red fabric, all the colors except red are absorbed and the leftover red is reflected to your eyes.

The rough white walls of the room consist of a kaleidoscope of tiny surfaces—like snow—that multiply lamplight into bright white. If you could pick out just one of these surfaces, you could see that it reflects an image just like an ordinary mirror. But the confusion of so many images melts into white. In the same way, a smooth piece of aluminum foil acts like a mirror until you crinkle it up; the more you crinkle it, the more it reflects like a white wall. This effect is most familiar on the slightly wrinkled surface of water, where the reflection of the moon or a streetlight is stretched into a long bright streak by the multiple reflections of parallel waves like so many lined-up mirrors.

During the day, our universal "lamp" is the sun. It spreads light as it reflects from treetops and clouds and bits of air. If there weren't any air for the light to reflect from, the sky would be black. Astronauts travel in the dark. Our night light is the moon, which reflects the light of the sun and also reflects the earth's own light back to us: This explains why you can often see "the old moon in the arms of the new"—the dimly glowing image of the former full moon nestled in the thin sliver of the next. The phases of the moon are reflections on the ever changing relationship among the sun, moon, and earth.

Reflection even made the atomic nucleus visible to Ernest Rutherford. It is still a major tool in unraveling the innermost secrets of the atom. When Rutherford bombarded a sheet of thin gold foil with subatomic particles, some of them bounced almost straight back. Today, people routinely use reflections of radio waves and sound waves—radar and sonar—the better to see all kinds of things. Earthquake waves reflected from the center of the planet help geologists to map its core. Sonograms bounce sound waves off internal organs, allowing expectant parents to peep in on newly developing babies only a few months old. Sound waves reflected from the ocean floor discover buried treasure and huge canyons miles beneath the sea.

One of the most enchanting things about reflections is their inherent symmetry, another kind of pervasive pattern. Mirror images are generally symmetrical. Yet symmetry to a scientist is not exactly the same thing as symmetry is to other people. To most people, a snowflake, for example, is highly symmetrical. But to a physicist, a billiard ball is the ultimate in symmetry. The higher the degree of symmetry, the more you can turn something through any angle and still have it look the same. Mirror images make pleasing symmetries because you can't tell the reflected image from the original object.

People are also highly symmetrical in many respects. We have right and left hands and feet, and right and left gloves and shoes to go with them. Yet we have hearts only on the left, and appendixes only on the right.

Things aren't always symmetrical in the nonliving, or physical, world, either. Until fairly recently, it was generally assumed that everything that happened in the physical universe *was* highly symmetrical; physics didn't discriminate between left and right. Then in 1957 Mme. Chien-shiung Wu discovered that there was a left/right difference in the way things decay radioactively. The question of symmetry in the physical universe is hardly answered. In many respects, the questioning has only begun.

The amount of symmetry you see in a situation, of course, can be highly subjective. If you are color-blind, the red light looks the same color as the green light, and you can't tell the difference between the command to stop and permission to go. You can't tell which way a boat is moving at night, because you can't distinguish the port (red) lights from the starboard (green).

Still, symmetry is a powerful and beautiful balance that pervades nature. For every left, there is a right; for every *yin*, a *yang*; for every color, a complement; for every particle, an antiparticle. Indeed, one way of looking at antiparticles is as "holes" in a virtual particle sea.

Indeed, "nothing" often displays a curious symmetry with "something." The two can be perfect complements. The part of the pattern that isn't present can tell you as much as the part that is. For example, mathematicians studying knots study knot complements—or "not knots"—to gain insights into the nature of these often convoluted forms. The "not knots" offer the same information as the knot itself in a different, perhaps more illuminating, perspective. In the same way, the missing places in the spectra of stars tell astronomers what elements make up the stars' atmospheres.

The artist Bob Miller is well known for the surprising things he does with "missing images," or shadows. In a piece of performance art that's become known as Bob's Light Walk, Bob lifts his hands high above his head and crosses the fingers of the two hands, so that the spaces in between them make a network of irregular holes. But there on the ground, the bright spots of light that get through the holes are perfectly round. They are shimmering images of the sun. The holes between his fingers serve the same purpose as the hole in a pinhole camera, or the pupil in your eye—selecting just enough light out of the surrounding confusion to make an image. These are the same sun images you see when sunlight filters through the irregular spaces between the leaves of a tree and scatters on the ground like so many spilled gold coins. During a solar eclipse, the round images slowly transform into crescents.

But if an image is a selected strand of information-bearing light rays, then what is a shadow? A shadow is a place where a similar strand of rays is blocked. Often Bob finds a fuzzy shadow on the ground underneath the tree and plucks a single sun image out of the air above it by making a single hole in his hand. And there on the pavement you can see a clear image of the sun with a dark tree branch crossing in front of it. But something truly remarkable happens when he holds up a small dark spot to cast a shadow, instead of a hole to catch an image. There on the ground is a single, round, dark shadow of the sun—and there crossing in front of the dark sun is a bright, white image of the tree branch.

A shadow, in other words, is a missing image. But it is also a complementary image, in the same way that a complementary color is left when you subtract one color from white—in the same way that night is the complement of day. It gives the same information as an image, but in a complementary form. In one of Bob's pieces, a pinhole shadow of a yellow sun, red house, and blue cloud shows up as a purple sun, green house, and orange cloud. Through a whole series of other sculptures and exhibits, Bob has shown that shadows can contain all the information in the light that casts the shadow in the first place. The shadow is every bit as rich in images as the light itself.

It's curious that shadows, like other "negatives," have such a shadowy reputation. After all, it is "negative" electricity that runs through wires and powers our homes and offices. This negative electricity is contained in a very real particle called an electron. It is no more or less real or less "positive" than its antimatter equivalent, a positive electron known as a positron. Indeed, what we call antimatter might be the everyday matter of some unknown universe. The only reason we call it anti- is because in one critical sense it is opposite from the matter that makes up ourselves. Making our peace with negatives seems especially important these days when even the vacuum is said to be teeming with energy, the universe emerged from "nothing," and the omnipresent computer stores as much information in the zeros of its digital code as in the ones. Nothing, it turns out, is very much of something. Or as the Greek philosopher Leucippus said, "*What is* is not more real than *what is not.*"

Even clear things, like lenses and prisms, cast shadows. If you hold up a lens or a pair of eyeglasses or a prism to a point of light (any nondiffuse light source will do), you can clearly see that it transports a bright image of the object from one place to another, leaving a shadow in its wake. When a prism spreads out white light into colors, it is really casting shadows. Each color appears in a place where the other colors are not; each color lies in the shadow of all the rest.

And light isn't the only thing that can be blocked to cast shadows. An umbrella or a building casts wind shadows and rain shadows as well as light shadows. A glass skylight casts a rain shadow while letting the visible sunlight through.

Few shadows block out *everything.* Shadows are really like filters. Rather than obscuring things, they select things, just as the coffee filter "shadows" the coffee from the grounds. Shadows let you filter the information from the noise, the important things from the distractions. Without them, you wouldn't be able to see the image for the glare. In fact, the screening or inhibition of extraneous nerve signals is as important to perception as the reception of those signals in the first place.

Shadows are created, of course, by obstacles. And often the shape of the shadow can tell you a lot about the obstacle that cast it. Aristotle looked at the round shadow of the earth on the moon during an eclipse and deduced that the world was a sphere. Film is the "obstacle" in front of the projector that creates the picture, just as your bones are the "obstacle" in the path of the X rays that create the X-ray pictures. Shadows also carry detailed information about the light falling on the obstacle and the surface on which it projects. Shadows that fall on curved surfaces are curved, and shadows that fall on curbs or stairs are zigzagged.

The spherical shape of the earth was also deduced from the fact that shadows falling on it at different places and times have different lengths and shapes.

Shadows are full of information. But in the end, they are only projections. Like the two-dimensional projections of the world that fall on our retinas, shadows lack depth and can be easily misinterpreted. The shadow of a cylinder can look like a circle or a rectangle or an oval, depending on how you play the light. This aspect of shadows allows us to use them to create wonderful shadow creatures out of the fingers of our hands. Like other two-dimensional images, shadows tell only one point of view. Part of the pattern is missing.

But then, that's one of the most interesting lessons modern physics teaches us about perception: Part of the pattern is always, *necessarily,* missing.

CHAPTER TWELVE

Order and Disorder

> The laws of nature might be as much the result of contingent and historical circumstances as they are reflections of some eternal, transcendent logic.
>
> —LEE SMOLIN

> The most extreme hope for science is that we will be able to trace the explanations of all natural phenomena to final laws *and* historical accidents.
>
> —STEVEN WEINBERG

CAN THE LAWS of nature allow for accident? Are the ubiquitous patterns of stuff and forces a matter of chance? How can the seemingly regular rhythms of space and time and matter evolve from (or dissolve into) churning chaos?

The strange interplay between order and disorder is one of those knotty problems that gives would-be natural philosophers (and real ones as well) serious headaches. Most of the world seems built on pattern, but even the most regular patterns dissipate into disorder with time. Conversely, the disorderly jiggling of stars and water molecules produces regular repeating patterns, like spiral galaxies and tornadoes. Order begets chaos, and the other way round. The predictable and the random change places. Accident directs the most deliberate events, and vice versa.

It's a tangled, but crucial, issue. Because for all the seeming order in the natural world—the crystals and snowflakes, light waves and spiral galaxies, tightly organized ant colonies and neat, elliptical planetary orbits—much of what we see around us is ruled by randomness. The shapes and sizes of stars are determined by the interplay between electrical pressure and gravity, but just why a certain star should happen to be born in just such a place at just such a time is all a matter of chance: A small random fluctuation in the configuration of molecules floating around in space can cause a few of them to clump together, making gravity a little stronger in that corner of the vacuum; more molecules and particles are attracted to this place, making the pull of gravity even stronger; eventually, even such a tiny chance trigger can turn out to be the underlying force behind the formation of an entire galaxy.

We owe our very existence to accidents. The air we breathe is a mistake made by ancient bacteria. A random mutation somewhere along the evolutionary line started the process of photosynthesis, which causes plants to breathe out oxygen. As a result of it, many plants died, poisoned by their own exhalations; those that lived created the sky. Plants themselves and indeed all living forms evolved from random combinations of atoms that happened to form large molecules suited for life; these organic ingredients of the early earth bubbled away for eons in their chaotic cauldron, buffeted by random collisions, energized by random lightning flashes, until some of them began to sparkle with signs of life. The infant steps of evolution have even been duplicated in laboratories: In August 1983, chemists announced that by sending an electric current through a soup of methane, nitrogen, and water they had managed to create "in one fell swoop" all the basic compounds that compose human genes.

Evolution into higher forms, of course, has been honed by natural selection. But the raw material for all such change is random variation. Every individual on earth is the result of the random mating of his or her parents, and countless other unaccountable events. "Whenever an infant is born," writes Loren Eiseley, "the dice, in the shape of genes and enzymes and the intangibles of chance environment, are being rolled again.... Each one of us is a statistical impossibility around which hover a million other lives that were never destined to be born—but who, nevertheless, are being unmanifest, a lurking potential in the dark storehouse of the void."

But what do we mean when we say that events are random? That things are ordered or disordered? That our lives are ruled by chance? We use the term *chance* in several different and quite contradictory ways. We say that something happens by chance in the sense of luck, or accident. This kind of chance is entirely unpredictable. On the other hand, we also use chance to mean *probability*—a way of predicting something. We predict a 40 percent "chance" of rain, or the odds of drawing a straight flush.

The meaning of order and disorder can be similarly confusing. What would you call a situation, for example, in which everything looked the same from all angles? In which everything was distributed evenly? Which was entirely undifferentiated and homogenous? A room full of coins that had exactly as many heads on the left half of the floor as on the right half of the floor? Or a universe where all the fundamental forces were the same and operated at the same strengths and distances? Or a classroom in which no distinction was made between the boys and the girls? Such a situation, it turns out, is both highly symmetrical and highly *disordered.* It is as if I took the entire contents of my file cabinet and flung them on the floor, so that there was no longer any difference between "Forces" and "Receipts," or "Gravity" and "Weisskopf." Disorder is completely democratic, but unspecialized.

Order, on the other hand, is much more authoritarian. A closet is well ordered if the shoes are separated from the shirts, and the skirts from the trousers. An army is well ordered if the privates are distinct from the colonels, just as an ant colony's order is based on the specialization of its member parts. Orderly situations are well differentiated. Today's universe with its four fundamental forces—with its atoms and organisms and galaxies—is far more ordered than the early universe, which was little more than a hot mass of homogeneous soup. Aristotle's universe was also highly ordered: Everyone from slaves to shoemakers, and everything from rocks to planets, had a proper and permanent place.

Order is obviously a lot more complex than disorder; chaos by comparison seem simple. A species is well ordered if its eating functions are well separated from its eliminating functions, for example, or if it has many other well-specialized parts. Curiously, we say we are a "higher order" of species, meaning that we are more complex and therefore clearly better than other species. But Darwin himself was careful not to attribute terms like "higher" and "lower" to the fruits of evolution. "For if an amoeba is as well adapted to its environment as we are to ours," writes Stephen Jay Gould, "who is to say that we are higher creatures?... Hair on a mammoth is not progressive in any cosmic sense." Only if it begins to get colder.

Physicists (as one might suspect) have even made orderly studies of disorder. Not only can they predict at least some aspects of random events; they have also discovered basic laws of nature in disorder. In fact, there is something eerily orderly about disorder. It is measurable, and even largely predictable. It is the one quantity in the whole universe that always, inevitably, increases. Like death and taxes, one thing of which you can always be certain is an increasing amount of large-scale disorder.

You obviously don't need to be a physicist to understand this. It came home to me (so to speak) one day when I was sitting in my kitchen contemplating all the frozen food that was going to spoil because my refrigerator had just broken. This came on top of the discovery that my right rear tooth needed root-canal work and that my son needed new sneakers. The garden was turning to weeds and my hair was turning gray. The house needed paint and the computer needed repairing. My best sweater was developing holes, and I was developing a deep sense of futility. After all, what was the point of spending half of Saturday at the Laundromat, if the clothes were dirty all over again the following Friday?

Disorder, alas, is the natural order of things in the universe. It is measured by a property that physicists call entropy, and the fact that it always increases emerges from the second law of thermodynamics. One way of stating it is this: "Natural processes tend to proceed toward a state of greater disorder."[13] Most fundamental physical quantities—such as energy, matter, momentum, and spin—are conserved; that is, you get out exactly what you put in, and the amount present in the universe always stays the same. You can't get rid of energy any more than you can create it out of whole cloth; you can only change it from one form to another.

But entropy is another thing. You always get *more* than you started with. Once it has been created, it can never be destroyed. The process is not reversible. The road to disorder is a one-way street. (The good news is that you can borrow energy from one part of the universe to create order in another part of the universe, thereby creating "islands of order," like stars and people. But more on that later.)

Because of its unnerving irreversibility, entropy is often called the arrow of time. Everyone understands this instinctively. Children's rooms, left on their own, tend to get messy, not neat. Wood rots, metal rusts, people wrinkle, and flowers wither. Even mountains wear down; even atoms decay. In the city, you see en-

---

[13]*Ideas of Physics,* by Douglas C. Giancoli (Harcourt Brace Jovanovich, 1978).

tropy in the run-down subways and worn-out sidewalks and torn-down buildings and fallen bridges. You know, without asking, what is "old." If you were suddenly to see the paint jump back onto an old building, you would know that something was terribly wrong. An egg does not unscramble itself any more than Humpty-Dumpty could put himself back together again.

But what is the cause of inevitably increasing entropy? What prevents Humpty-Dumpty from spontaneously reassembling? Or for that matter, people from growing younger instead of older? The answer is essentially *probability:* the combined results of countless random events. "Irreversibility," Richard Feynman concludes, "is caused by the general accidents of life."

Take the air in my kitchen just before the refrigerator broke. It was a highly ordered situation—a low degree of entropy. All the cold air was kept inside the refrigerator, and the warmer air was isolated outside. The minute the machine stopped working, however, the cold- and warm-air molecules were free to exchange energy at random. As they were jostled about this way and that, it was possible, of course, that all the cold (meaning slow-moving) molecules would be bumped back in the direction of the refrigerator. But it would be highly unlikely. The likely result was that the cold and warm molecules would wind up randomly mixed, and that I would be left with a lukewarm mess.

Of course, there was nothing to prevent any one molecule from moving one way or another. No force pushed the cooler molecules away from the refrigerator. In fact, any one of the slow-moving molecules stood about as much chance of being bounced back toward the refrigerator as of being bounced away from it. But take trillions and trillions of warm and cold molecules mixed together, and the chances that all the cold ones will wander toward the refrigerator and all the warm ones will wander away from it are practically nil.

Entropy wins not because order is impossible but because

there are always so many more paths toward disorder than toward order. There are so many more different ways to do a sloppy job than a good one, so many more ways to make a mess than to clear it up. If I put a baby at my computer keyboard, the odds are just over one in a hundred that she will type the letter *a.* There is less than a chance in a million, however, that she would consecutively strike the letters that spell out "Shakespeare." And the chance is so infinitesimal that she would type the complete works of Shakespeare that we call it impossible.

This is precisely the same reason that it is "impossible" for warm air to randomly excite the molecules in a melted stick of butter so that they regroup spontaneously into a bar—or for ice cubes to spontaneously appear in a lukewarm drink. Because there are far more opportunities for things to turn out otherwise. A baby learns to take a puzzle apart long before he learns to put it back together again, because there are so many more ways for the puzzle to come apart than there are for it to go together. There is only one possible way for Humpty-Dumpty to remain a recognizable whole, but there are an infinite number of ways for him to fall to pieces. In fact, this explains why most random mutations are harmful rather than helpful: There are simply many more ways that a random change in the nature of things will turn out for the worse.

The more pieces there are in the baby's puzzle, the harder it is to put it together. In a sense, entropy boils down to the number of possibilities. A coin can come up only heads or tails. But a dust particle in a room can take an almost infinite number of possible positions and so add up to a far "messier" situation. If my kitchen contained only a dozen or so air molecules, then it would be likely—if I waited a year or so—that at some point the six coldest ones would congregate in the freezer. But the more molecules in the kitchen—the more factors in the equation—the less likely it is that their paths will coincide in an orderly way. Or as one physicist put it, "Irreversibility is the price we pay for complexity." That's how we can have an energy crisis even though energy itself is always conserved. While the total amount of energy stays the same, useful energy is another matter entirely. Once it's gone, it's lost forever. The second law of thermodynamics says that no matter how efficient you make a machine, you always get out less useful energy than you put in. The excess energy gets dissipated into heat, or entropy. And that energy can never be retrieved. Once the water has flowed over the falls, it has lost its potential for useful work. Once the cold-air molecules had been let out of my freezer, they lost their potential to perform a useful function; they allowed my butter to melt, my milk to spoil, my frozen vegetables to decay. "What has been 'lost' in the irreversible process is not *energy,*" says Feynman, "but *opportunity.*"

There is an inherent contradiction in all this, however. On the one hand, physicists say that an increase in disorder is inevitable—and you can see the results of this around you wherever you look. On the other hand, the universe is clearly an increasingly structured place—and the results of this are obvious, too. The primordial ball of fire has cooled and condensed from a hot, amorphous mass into elements, stars, planets, people. What we see in the universe is increasing order, not disorder.

The paradox becomes compounded when you consider the very close relationship between disorder and temperature. Heat is a measure of random motion. Warmth (as in warm, melting butter) means disorder. The early universe was a disorderly hot mess of particles and radiation. Today even the stars are made not of atoms but of something more like an undifferentiated stew of atomic particles—pieces of atoms torn apart by the high energies inherent in extremely hot temperatures. This state of matter is called a plasma, which means "mixture," and it accounts for all but a fraction of the matter in the universe.

Cold, on the other hand, is associated with order. Only when temperatures cool considerably can stable atoms form out of atomic constituents like protons, neutrons, and electrons. Only when they cool further still can more complex molecules form from the atoms. A molecule of water breaks down into hydrogen and oxygen atoms if things get too hot. Steam is less orderly than water, and water is less orderly than ice. All the familiar "orderly" states of water—ice cubes, snowflakes, hailstones, crystals—can form only at comparatively low temperatures. The increasing order in the universe is also associated with a cosmic cooling down. The billion-plus-degree temperatures present at the creation—the so-called Big Bang—are now measured at a mere three degrees above absolute zero. Physicists say that the four fundamental forces have somehow "frozen" into their now separate states.

But how can the universe be cooling down and becoming more ordered when entropy is universally increasing?

The answer is that you pay for order with energy. It takes energy to produce a neat set of files or a well-ordered closet, just as it takes energy to produce an atom or a star or to keep all the cold-air molecules inside the refrigerator. It *is* possible to create order in the universe, but the energy to do it must be borrowed from other parts of the universe. The islands of order that bask in the chaos—the crystals and snowflakes, the buildings and cities—all exist at the expense of something else. We get the energy to construct buildings, for example, primarily from the fossil fuels needed to fire the steel mills and drive the cranes and trucks; in doing so, we increase the familiar form of entropy known as smog. The price of creating order in any one corner of the universe is increasing disorder somewhere else.

When it comes to the whole universe, some physicists explain that the increase in disorder dissipates into the vastness of infinite space. After all, if entropy can be measured as the number of possibilities, then the number of possibilities in infinite space is clearly limitless. Others say it sinks into the gravity field, or black holes. But it has to go somewhere.

The most obvious exception to entropy is life. A seed soaks up some soil and some carbon and some sunshine and arranges it into a rose. A seed in the womb takes some oxygen and pizza and milk and transforms it into a baby. Death is an extreme form of entropy. Life is the epitome of order, of purpose incarnate. To live is to be in constant battle with the second law of thermodynamics, and yet it is a battle more often won, it seems, than lost. Despite all odds, flowers bloom in the desert and children bloom in slums. "If something or somebody has a will to live, you see," says Murchie, "it or he [or she] must resist diffusion and move from disorder to order, which means avoiding all those easy paths away from the previous position in favor of returning to it, or, better, staying with it from the beginning. Basically it connotes sticking around."

The inevitable accidents and obstacles of life almost guarantee that things will get off track, bounced onto random paths. Disorder is the path of least resistance—the easy, but not the inevitable road. Social institutions, like atoms and stars, decay if energy is not expended to keep them ordered. And every time entropy increases, more opportunities are lost to stop the avalanche of disorder that threatens the physical and social universe alike.

CHAPTER THIRTEEN

Cause and Effect

A thing may be indeterminable but not indeterminate. Nature knows what she is doing, and does it, even when we cannot find out.

—SIR ARTHUR EDDINGTON

Natural philosophy consists in discovering the frame and operations of nature... and thus deducing the causes and effects of things.

—ISAAC NEWTON

EVEN MORE convoluted than the relationship between order and disorder is the uneasy pairing of cause and effect. Nothing seems more important to physicists and philosophers than pinning down the causes of things. The Greek philosopher Democritus said he would rather understand one cause than be king of Persia. And no wonder: It is one thing to know *how* the universe works and quite another to know *why* it works that way. People understand how gravity works but not why, in the same way as they can explain to you how schoolchildren murder their classmates but are at a loss to tell you why.

One of the great unanswered whys in history concerns the unexplained astronomical amnesia that many historians say descended on the Western world between the time of the ancient Greeks and the Renaissance. Aristarchus of Samos was writing in the third century B.C. that the spinning earth and other planets revolved around a central sun, yet it was eighteen centuries later before this "discovery" was resurrected by Copernicus and others. "We know how this happened," writes Arthur Koestler. "If we knew exactly why it happened, we would probably have the remedy to the ills of our own time."

The search for underlying causes is innately and powerfully appealing—not the least because it implies the ability to control: If you know what makes things happen (or not happen), you might be able to make them happen (or not happen) again. Above all, people like to think that there *are* causes. It is disconcerting to think that events are random, that the things we see around us do not conform to an understandable (if still not understood) body of natural law. Newton's ideas were readily accepted in part because they contained a clearly applied formula for cause and effect: Planets orbited and falling objects accelerated in a certain way because gravity pulled on them in a certain way. Darwin's ideas are much harder for some to swallow, and at least some of this resistance is rooted in the role his theory gives to random mutations. (Astronomer Sir John Herschel complained that Darwin's ideas about evolution constituted little more than "the law of higgledy piggledy.") Even Einstein refused to accept quantum mechanics because he thought it did away with causality altogether. He wrote to his friend and fellow physicist Max Born in 1944, "You believe in the dice-playing god, and I in the perfect rule of law in a world of something objectively existing which I try to catch in a wildly speculative way. I hope that somebody will find a more realistic way, or a more tangible foundation for such a conception than that which is given to me. The great initial success of quantum theory cannot convert me to believe in that fundamental game of dice."

For Einstein and others, quantum mechanics seemed to introduce an unacceptable degree of uncertainty and unpredictability into the universe. Perhaps not since Copernicus finally put the sun at the center of the solar system has science so profoundly ruffled people's philosophical feathers.

A strict belief in causality, on the other hand, seems to pull the rug out from under the possibility of free will. If every effect is produced by a cause that itself is an effect produced by yet another cause, and so on in a straight line that takes us back to the beginnings of the universe, then everything we do must be predetermined—and must have been since the beginning of time. If the cause of your missing the train is a snowstorm that was caused by a warm front over the Atlantic Ocean two weeks ago, which itself was caused by a particular combination of winds and sunspots and so on, you could easily trace the "cause" back to any arbitrary event you wished—an event that would still have a cause. Believing in this "endless chain of natural causes," argued Born, "leads necessarily to the idea that the world is an automaton of which we ourselves are only little cogwheels. . . . I cannot enlarge on the difficulties to which this idea leads if considered from the standpoint of ethical responsibility."

Newton's universe was such an endless chain of natural causes: The behavior of planets and presumably people could be calculated if only one had enough information and enough time. Everything boiled down to pieces of matter propelled and controlled by precise and predictable motions. Emotions and thoughts were merely manifestations of so much preset electronic circuitry. The illusion of "free will" was nothing more than the arrangement of atoms and molecules that made up the human body. "Exhorting a man to be moral or useful," said Sir James Jeans, "was like exhorting a clock to keep good time; even if it had a mind, its hands would not move as its mind wished, but as the already fixed arrangement of its weight and pendulum directed."

It's all too tempting to link two events in time and then say that one was caused by the other. Johannes Kepler's mother was arrested as a witch, in part because her visit to a neighbor's house unhappily coincided with the onset of a major illness. No educated person today would make this inference. But you do hear people say that welfare causes poverty, or that family planning causes teenage pregnancies.

Even when things are clearly and consistently connected, it is not always possible to make causal connections. Low-flying swallows do not cause rain, any more than orthodontia causes puberty. As Born points out, "You can predict (with the help of a railway timetable) the arrival at King's Cross of the ten o'clock from Waverly; but you can hardly say that the timetable reveals a cause for this event."

Confusion about the relationship between cause and effect frequently leads to amusing (in hindsight) conclusions. For example, Guy Murchie tells of the committee of scientists who decided not to go ahead with a large investment in Gutenberg's printing press, because, they said, there could never be a big enough demand for books. The reason? Only 1 percent of the population could read. It probably never occurred to the scientists that the availability of books could be a *cause* of people learning to read as well as vice versa.

This very elusiveness of causes is what makes them such fertile territory for misinterpretation and superstition. The truth is that we rarely understand what forces are at work. What causes a family quarrel? A rising or falling crime rate? The crash of a jetliner? Most of the time it's impossible to untangle the cause from the complex of circumstances that surrounds it. In many ways, the cause *is* that complex of circumstances—and it may include everything from the day's weather to the full weight of history. You might as well ask, What causes matter? Or what causes life? Or, as Victor Weisskopf once said to me, "What is a cause? All you can say is, there are connections."

Connections between seemingly related events have been assigned to an ever changing cast of causes throughout history. Ancient people didn't understand the connections behind spring and fall, or day and night, so they attributed these things (along with almost everything else) to the will of various gods. The gods were kept very busy, for there was a lot to do: not only the detailed arrangements of floods and famines but also the everyday events of everyone's life. Nothing happened as a result of consistent causes or understandable reasons. Even the planets would have stopped moving if the angels pushing them had stopped fluttering their wings. In Aristotle's day, it took no less than fifty-five separate spirits just to keep the planets in motion.

Kepler, according to most sources, was the first person to seriously speculate, in the sixteenth century, that some "force" must be involved in the motions of the planets; he was the first to propose that things happen for rational, nonmystical reasons. Newton crystallized this kind of thinking in his vision of a great clockwork universe, where everything was controlled by forces. "For the eighteenth century, the world was a giant mechanism," wrote J. Robert Oppenheimer. All motions could be analyzed in terms of the forces producing them. In a sense, Newton substituted one kind of cause and effect for another. Forces took the place of spirits and gods. The worldview of causes went from chaos to order, from total randomness to complete predictability.

Now it seems as though we have come full cycle. Quantum mechanics, with its innate uncertainties, has been accused of reopening the door to randomness, dispensing with order and causality, distilling the laws of nature into a kind of subjective mysticism. Or at least so say some popular interpretations. In truth, all that quantum mechanics has done is to bring to the fore a new *kind* of cause. And surprising though it may seem, there are many *kinds* of causes.

In the first place, there is the kind of cause that says things happen because it is the natural order of things. Copernicus, for one, thought that gravity was just a natural inclination of matter, bestowed by the Creator. Rocks "belonged" down, just as clouds "belonged" up. "The desire of every heavy body is that its center may be the center of the earth," wrote Leonardo da Vinci. Yet this kind of thinking goes clear back to Plato and Aristotle, who thought that slavery was the natural order of things just as they thought that circular motion was the only "natural" orbit for a planet. Aristotle's was a multilayered universe where everything had its proper place; a "cosmos graded like the Civil Service," Koestler called it. Aristotle's influence lasted a millennium and a half, and still persists with those who argue that some kinds of people are "naturally" poor or stupid, or that women "belong" in the kitchen. It is based on the same kind of thinking that says smoke has a "natural" tendency to rise or that the sun "belongs" in the center of the solar system. It is largely an admission that we do not know the forces or reasons involved.

Of course, our notions of what's natural do change, even in physics. Aristotle thought that the natural state of things was at rest and that you needed a force to keep them going; Newton said that the natural state of things could be in motion and that the "force" that kept them going was a "natural" property of all matter known as inertia.[14] Newton thought it "natural" that there was such a thing as absolute motion and absolute rest, but Einstein proved him wrong.

Natural causes of the sort that Aristotle had in mind are very different from the causes we associate with forces. A force

---

[14]Newton's view was "righter" than Aristotle's mainly because it resulted in progress. See "Right and Wrong."

requires an exchange of energy. You punch someone and he falls down. You blow on a birthday candle and it goes out. You wave a greeting and send a subtle flux of air molecules into the surrounding space. Gravity pulls and an apple falls to the ground. And yet even gravity remains essentially a name for a pattern that is associated with a still-not-very-well-understood behavior of objects. Newton never claimed to understand why gravity worked, or how it spread through space. "I do not deal in conjectures," he said. Or as Richard Feynman put it, "At the time of Kepler, some people answered this problem [of what makes the planets go around the sun] by saying that there were angels behind them beating their wings and pushing the planets around an orbit. As you will see, the answer is not very far from the truth. The only difference is that the angels sit in a different direction and their wings push inwards." (The inward push of the angels, of course, is gravity.)

The notion that forces are carried by fields, or exert their influence on bodies by means of fields, seems uncannily connected to Aristotle's notion that everything has its proper place in the cosmos. That is, the apple falls not because gravity pulls on it but because the apple naturally falls to its proper place in the gravitational field, just as iron filings naturally fall into place around a magnet. In fact, the tendency to fall into a natural state is common to all things in nature. Not only do apples fall to their ground states but so do atoms—in the process giving off energy and emitting light. All things seek their lowest energy state, just as water seeks its lowest level. And this tendency to seek the lowest or most stable level is considered a perfectly legitimate *cause* of things.

Yet there are important differences between this kind of cause and the causes that operated in Aristotle's universe. The natural states of apples and atoms are consistent, not capricious. Gravitational fields behave the same way toward men and women, apples and oranges. Atoms change states whether or not they have made penance to a particular god. Forces and fields do not discriminate. They treat the atoms in earth, fire, air, and water equally.

This consistency has its roots in a peculiar kind of "cause"—symmetry. All the strange effects of relativity, from time dilation to curved space, result from the idea that the laws of nature are symmetrical; that it does not matter whether you are moving or at rest; that the speed of light is always the same; that there is no absolute rest frame in the universe. What is the "cause" that makes time slow down as you travel at greater speeds? Why do electrons traveling at 99.999 percent the speed of light in giant accelerators gain forty thousand times their normal weight, or mass? The "cause" is symmetry: the fact that the speed of light is measured the same no matter how fast you are moving; that some differences do not make a difference. The reason that rulers can change their dimensions and clocks can tell different times is that you always see the *same* laws at work in the universe, no matter how you are moving about in it.

Like other causes, the idea that symmetry can be a shaper of things goes back to ancient Greece. Plato said that the shape of the world must be a perfect sphere, and that all the planets must travel in perfect circles because only circles are perfectly symmetrical. Circular motion has no beginning and no end. It doesn't matter how you look at a circle, it always looks the same. Some people even argued that objects "gravitated" toward the center of the earth because that made things nice and symmetrical. The notion of symmetry was such an appealing "cause" that it wasn't until Kepler that people officially recognized that the planets orbited not in circles at all, but in ellipses.

My friend the physicist liked to point out that symmetry is also the basic argument behind social ideas such as the need for civil rights legislation: People are innately symmetrical before the law. You can't treat blacks and whites (or men and women) differently, because their fundamental needs and abilities are the same. In fact, while it may be easier to think of causes in terms of physical forces ("after all," says Jeans, "our hairy ancestors had to think more about muscular force than about perfect circles or geodesics"), no concept of cause that relies on force can be entirely subjective. According to relativity theory, force itself depends on the observer's motion and point of view. So causes interpreted as forces must always be necessarily somewhat subjective, too. Looking for causes in symmetries leads to much clearer and more uniform understanding.

Only in quantum mechanics—in the physics of subatomic things—do causes seem to spring from nowhere. Even the whims of the ancient gods seemed more understandable than the inner workings of the atom. Yet there is a magical order to the seeming randomness of atomic events. And the very reality of the order in random events has totally altered the meaning of our notion of causes.

What is the "cause," after all, that makes a tossed coin turn up heads 50 percent of the time and tails 50 percent of the time? What is the cause that makes a certain number of radioactive atoms decay now instead of later? What is the cause that determines how many times the roulette wheel will come up red or black? It is the same kind of cause that keeps an egg from unscrambling itself, yet makes it almost certain that a room will unclean itself; the same cause that makes the heat flow from the warm drink into the melting ice cube and not vice versa. The cause that makes these things happen is simply that they are more likely to happen than the alternative. A cause, according to my friend the physicist, is anything that makes something happen at a slightly higher rate. Those things that happen more often are those things that have the most ways of happening, which is another way of saying that most probable things happen more often. So probabilities can also be causes. This seems completely nonsensical only until you stop to consider the evidence.

Take a bunch of billiard balls, for example. (Physicists are always taking a bunch of billiard balls.) Many science museums have an exhibit in which billiard balls are allowed to fall at random through a forest of pegs sticking out from a wall. The balls hit the pegs and bounce this way and that, all at random. And yet when they all collect at the bottom in a bin, they take the shape of a surprisingly predictable curve. People like to try this experiment again and again, precisely because the result seems so unlikely. How do you get that nice pattern out of all that random motion? What causes it?

Or take a drunk leaning on a lamppost. Say he decides to go for a walk. First he steps forward, then teeters sideways, then stumbles backward, then goes off in another direction—all at random. Can you predict how far the drunk will travel after taking a given number of steps? Incredibly, it turns out that you can. The total distance traveled equals the average length of each straight step times the square root of the number of steps he takes. So if his average step is one yard long, he'll travel ten yards in one hundred steps. (You only don't know which *direction* the steps will take him.)

Using this same kind of analysis, Einstein calculated the size of molecules from looking at the chaos of microscopic collisions known as Brownian motion. Small particles suspended in a liquid—say, plant spores or oil droplets—are buffeted about at random by the unseen molecules of the liquid. The result is that they move about somewhat like drunks, and their paths can be calculated in much the same way. The same equation is used to predict how fast smoke—or pollution—will spread in the sky.

Isaac Asimov even got the idea for his *Foundation* series of science fiction stories from the statistical order that emerges from the random motions of gas molecules. Given quintillions of quintillions of molecules, he says, you can predict exactly how the sample will act:

> The motion of any one atom or molecule is completely unpredictable—you can't tell where, in which direction, or how fast it will move—but you can average all the motions and from this deduce the gas laws. It occurred to me that this might also be true for human beings. A human by himself is quite unpredictable, but a mob is usually a little more predictable. What if we applied a kinetic theory of humans to some future time, when there were millions of planets full of people?

As Asimov points out, this is only science fiction. People are much more complicated than gas molecules. Yet the idea of statistical probability takes on a strange concreteness when it comes to subatomic particles: All particles can also be described as waves; this is part of the dual nature of matter. But the particles can also be pointlike, because the waves are not matter waves; they are probability waves. The wave charts the probability that the particle will be in a particular place at a particular time if you should choose to measure it.

People often make the mistake of dismissing probabilities as mere abstractions, and therefore unreal. They do not take them seriously as causes. Yet the high probability of, say, a natural disaster or nuclear war can cause it to happen as much as a high probability causes a coin to land tails 50 percent of the time. Probable causes, like probability waves, are real because they *work.* The laws of statistics are laws of nature like any others. Probability is as much a shaper of things as gravity.

Probable causes also introduce the paradox that small quantities appear to obey different laws of nature than large quantities. The behavior of one coin or atom is completely unpredictable, while the behavior of hundreds of coins or trillions of atoms is quite precisely predictable. Unpredictability implies randomness, which is equated with lack of cause. Either something happens on purpose (with a cause) or by accident (at random); you can't have it both ways. The behavior of atoms, according to this interpretation, is random and therefore acausal.

What causes a radioactive atom to decay, for example? Say you take a milligram of radium. You can predict fairly precisely how many of its atoms will disintegrate with every passing second. There is absolutely nothing you can do to change this situation. The rate of decay is not affected by anything in the environment. You can make it hotter or colder. You can alter the motion, or crowd the atoms closer together, and the rate will stay the same. On the other hand, nothing in the past history of the atoms determines what they will do, either. Any milligram of radium from anywhere in the universe will behave exactly the same. There is no way, internally or externally, to change the situation. There is no determining factor in the past or in the present. Therefore radioactive decay seems truly to be an event without a cause. And, in fact, almost anything that has to do with a single atom exhibits this same "acausal" property.

Yet "how could this be," asks physicist John Wheeler, and "leave the largely familiar world intact as we know it? Large bodies are, of course, made up of atoms. How could causality for bullets and machines and planets come out of acausal atomic behavior? How could trajectories, orbits, velocities, accelerations, and positions reemerge from this strange talk of states, transitions, and probabilities?"

If God plays dice with the universe, then he (or she) presumably plays a great deal of dice; otherwise, how to account for the familiar and predictable laws of nature?

In the second place, we are left with the further and perhaps more fundamental paradox that *chance* follows *laws!* And that events ruled by cause and effect, on the other hand, are seldom precisely predictable! It seems, as Born writes, "a hopeless tangle of ideas."

Our ability (or inability) to predict something does not necessarily depend on our understanding of causes—something that becomes evident when you consider how many things we can predict even though we do not understand them (low-flying swallows predict rain), and the many other things we cannot predict even though we *do* understand them. Take weather, for example. The forces behind weather are well understood. But weather itself is highly unpredictable, primarily because it is so complicated. You probably can understand and predict the course of a single air molecule as it responds to changes in humidity and atmospheric pressure. But take a whole bunch of air molecules and you are lost.

As in foreign affairs and personal affairs, one small unseen effect can be enough to change the entire configuration of events. Nature (including human nature) is often too complicated and interconnected for neat categorizations into cause and effect.

So did the innate uncertainty in quantum mechanics do away with causality or didn't it? Is the universe at its core a precisely tuned clockwork mechanism? Or is it random—a room full of tossed coins? Summing up the fruits of quantum physics, J. Robert Oppenheimer wrote, "We saw in the very heart of the physical world an end of that complete causality which had seemed so inherent a feature of Newtonian physics." But Max Born concludes, "The statement, frequently made, that modern physics has given up causality is entirely unfounded.... Scientific work will always be the search for causal interdependence of phenomena."

It may be that the only thing that has been lost by opening the Pandora's box of quantum physics is the assumption that understanding causes also means the ability to predict and control things. As Weisskopf points out, you still know that a radioactive atom will decay and you still know how it will decay—"you only don't know when." In fact, the crux of the uncertainty principle boils down to a matter of timing. Because the more accurately you try to determine exactly *when* something will happen, the more you make other factors obscure. On the other hand, there is a sense in which time itself becomes a cause of things. "Give me a million years," writes Stephen Jay Gould, "and I'll flip a hundred heads in a row more than once." When it comes to evolution, "time is in fact the hero of the plot. Given two billion years or so, the impossible becomes possible, the possible probable, and the probable virtually certain."

You don't need to know every factor in the causal equation, that is, to determine the probable outcome. You know that given a certain number of handguns in general circulation, a certain speed limit, a certain population of smokers or poverty rate, that a certain number of people will die. You only don't know who. But you can certainly say that "there are connections."

In the end, Born harks back to the idea of complementarity. Strict causality and absolute randomness both have their place in the scheme of things, but both together are as inconsistent as images of waves and particles. In fact, if you take both arguments to their logical (linear) conclusions, they make no sense at all. For if you say that any kind of cause at all determines the way things (or people) behave, then you have to come to the conclusion that everything is predetermined. On the other hand, if you say that nothing determines the way things behave, then you must conclude that everything is random. If there are causes for everything, then we are cogs in the clockwork. If there are not causes, then we are so many dice.

On the surface, causality and randomness may seem to be mutually exclusive, but on closer inspection they must be seen as complementary facets of a larger reality.

CHAPTER FOURTEEN

Small Differences

Blobs, spots, specks, smudges, cracks, defects, mistakes, accidents, exceptions, and irregularities are the windows to other worlds.

—Artist's statement, by BOB MILLER of San Francisco

The existence of stars rests on several delicate balances between the different forces in nature.... In many cases, a small turn of the dial in one direction or another results in a world not only without stars, but with much less structure than our universe.

—LEE SMOLIN

PEOPLE SAY THAT little things mean a lot. While not always true, it is surprising how many large-scale phenomena seem to be ruled by the tiniest incremental changes, how many major discoveries have been cracked by a scientist who noticed an almost imperceptible anomaly.

The existence of antimatter was first signaled when physicist Paul Dirac stumbled upon an errant minus sign in an equation. The principle of the electric motor was discovered during a classroom demonstration when high school teacher Hans Christian Oersted happened to notice an unexpected deflection of a current-carrying wire as it moved through a magnetic field. The planet Neptune was found as astronomers tried to account for a small irregularity in the orbit of its neighbor, Uranus. Einstein's theory of general relativity was first "proved" by a deflection of starlight passing near the sun by a mere 1.75 seconds of arc (less than a thirtieth of a degree).

More recently, the first planets orbiting distant suns were discovered as slight wobbles in the positions of the parent stars; the mass of a particle known as "the spinning nothing"—the nearly nonexistent neutrino—was discovered (perhaps) as a rare streak of light in a 12.5-million-gallon water tank lined with photomultiplier tubes more than 3,250 feet beneath the Japanese alps. If the discovery bears out, the little neutrino may well make up much of the mass of the universe.

One obvious reason that small differences make a big difference in scientific understanding is that small differences can make a big difference in nature. A single extra electron in an atom's outer shell spells the difference between sodium—one of the most chemically active metals—and neon, a chemically inactive gas. An extra neutron turns plutonium-238, good for powering spacecraft but extremely hazardous to human health, into plutonium-239, used for making bombs. Plutonium-239 splits easily, setting off chain reactions. Plutonium-238 does not split so easily, but spits out radiation at an enormous rate; if it lodges in someone's lungs, it can be a death sentence.

Other examples abound: A minute change in the wavelength of light turns blue to violet, and allows violet, but not ultraviolet, to pass through glass. If the "strong" force were a little weaker, or the electric force a little stronger, atoms—and therefore matter as we know it—would probably not exist. Indeed, the existence of matter at all is based on what must have been a small imbalance of matter and antimatter sometime near the earliest moments of the universe: Particles of matter and antimatter annihilate in a burst of energy when they meet; if the universe had been created with equal parts of both, then pairs of particles and antiparticles would have annihilated long ago, leaving nothing but radiation.

Living things are even more profoundly affected by small differences. If the earth had settled into an orbit a little closer to the sun, temperatures would have been so hot that organic molecules could not stick together. If the orbit had been a little farther from the sun, colder temperatures would have frozen opportunities for life into immobility. A tiny alteration in the structure of DNA can mark the difference between brown eyes and blue, between sickness and health, between extinction or survival of a species. One geneticist estimated that the difference between a mild virus and a killer can be as small as three atoms out of more than 5 million.

These minor differences between major groups carry endless fascination. The writer Annie Dillard points out, for example, that the difference between the lifeblood of plants and people is but one atom: Chlorophyll is made up of 136 atoms of hydrogen, carbon, oxygen, and nitrogen arranged in a ring around a single atom of magnesium; hemoglobin (blood) is made up of 136 atoms of hydrogen, carbon, oxygen, and nitrogen arranged in a ring around a single atom of iron. The genetic difference between humans and African great apes is less than 1 percent.

Of course, some small differences make much more difference than others do. If every small difference threw things askew, then the universe would be hopelessly unstable and the forms of life on earth would change abruptly every time you turned around. In truth, most small differences don't make much of a difference. Slight changes in temperature, in the shape of your nose, in the way a sentence or a species is constructed, can hardly matter at all. But if the small difference pops up at a crucial point, then it can make all the difference. The difference between 98.6 degrees F and 106 degrees F is both small and potentially lethal.

Sometimes the all-important small difference seems like icing on the cake, when in reality it is the essential ingredient that *makes* the cake. The difference between an amateur flutist and Jean-Pierre Rampal might not even be noticeable to the eye. But it would be huge to the ear. The same is true of the last millisecond of time that turns the runner into a champion, the almost imperceptible turn of phrase that gives power to the poem, or the embellishments of style and grace that turn a good dancer into a great one. As philosopher Yehoshua Bar-Hillel said, "The step from not being able to do something at all to being able to do it a little bit is much smaller than the next step—being able to do it well."

Small differences usually make big differences when things are connected, like rows of dominoes. A rock doesn't start a landslide unless a whole mountain of rocks is somewhat shaky and ready to fall. When it is, the first rock acts like a trigger, setting off a gun that is already loaded and cocked. The trigger could be the combination of genes that sets off a cancerous growth, or the push on the first domino that knocks down the row; the ice crystals that set off a chain of events that culminates in a hurricane, or the small crack beneath the surface of the earth that finally causes Mount Saint Helens to blow its stack.

Unfortunately, it is not always easy to know which things are connected—and how. That is why earthquake prediction, for example, is so difficult. Seismologists have yet to discover exactly which of the myriad of strains and stresses near a fault will cause two plates of the earth's crust, locked together by friction for years, to lurch apart in an earthquake. So many small things seem to be important. The same is true of human psychology. No one can predict which combinations of small events will erupt in a murder or suicide.

Some small differences, however, are connected in such a way that the results are literally explosive. If you knock down one domino in a row, the chain of dominoes passes along the effect and at the end of the line you have one knocked-down domino (with a line of other knocked-down dominoes in between). But say you have a series of dominoes set up in such a way that the first domino knocks down two dominoes, and those two knock down four, and those four knock down eight, and so on. Then you have what is known as a "nonlinear sequence of events." At the end of the chain, that single first domino may fell hundreds of dominoes. The effect is not commensurate with the cause, any more than anger at rudeness on the road is commensurate with murder. What you have, in effect, is an explosion.

What we normally think of as small differences—for example, multiplying something by a tiny number such as 2—leads to everything from epidemics to the nuclear fusion that powers the sun. Doubling once or twice or even three times doesn't necessarily make a big difference. But if you double anything—no matter how small—enough times (even a relatively small number of times), you always end up with a huge amount. Because doublings are connected like dominoes, each additional doubling is connected to all the doublings before. For this reason, doubling the thickness of a piece of tissue paper just fifty times adds up to enough mileage to reach the moon and back seventeen times.

Yet as important as these small differences are, they can be wickedly hard to perceive. Most people don't notice when the department store charge account adds 1.5 percent interest to their balance each month, and it's hard to get excited when you read that the world population is increasing at a rate of 1.8 percent. The differences themselves seem small. It's the *connections* between the differences that make them add up—the total pattern—and it's the total pattern that we can't perceive.

Several years ago, physicist Albert Bartlett described this situation vividly in the *American Journal of Physics.* Bartlett took a population of bacteria doubling inside a Coke bottle once per minute. They started at 11:00 A.M., and by noon the bottle was full. What time would it be, Bartlett asked, when even the most foresighted of the bacteria realized that they were running out of room? The answer: 11:58 A.M. And even then, the bottle would still be *three-quarters empty*—so they would have to have been very foresighted bacteria indeed. At 11:59 A.M., the bottle would still be half full—or half empty, depending upon your point of view. No doubt, said Bartlett, the presidents of the bacteria bottle companies would be running around Bacterialand assuring everyone that there was no reason to limit the growth rate because, after all, there was more room still left than had ever been used in the population's entire history. And then—just suppose—they mounted an extensive effort to explore for new space offshore and, lo and behold, found three new Coke bottles! All the space-starved bacteria would breathe a long sigh of relief. But how much time would they have before they were out of room again? Answer: two more minutes (two more "doubling times").

An instructive (although somewhat different) example comes from a fable reportedly told by Jorge Luis Borges. It concerns a somewhat mad prince who built columns throughout a vast park. Each one was a beautiful bright red. And each was just like the first. As you passed along, you would see dozens and dozens of them, all the same. And then all of a sudden after a week or so of traveling you would realize that the bright red columns had mysteriously metamorphosed into pure white pillars. How could it be? The answer is that the difference in color between one pillar and the next was too small to discern. But the accumulated difference was enough to change bright red to white. The story is based on the same insight that moved British statesman Edmund Burke to remark in the late eighteenth century: "Though no man can draw a stroke between the confines of day and night, yet light and darkness are upon the whole tolerably distinguishable."

In an entirely different context, the effect of small differences can explain why imperceptible racial or gender bias—bias that no one can quite put a finger on—can lead to widespread discrimination. Hunter College psychologist Virginia Valian described this process in an interview with *New York Times* science writer Natalie Angier:

> People are often unable to perceive or assess how small imbalances can really add up. One computer simulation showed what happens in an organization with eight levels of hierarchy, when it is staffed initially with equally numbers of males and females, but the men are given a tiny advantage in the promotion process—a mere 1 percent. The programmers continued the simulation until there had been a complete staff turnover at their fictitious organization. At the end, the top level was 65 percent male, 35 percent female.
>
> Any single instance of bias is likely to be tiny, and someone might say, "You're making a mountain out of a molehill." But mountains are molehills piled one on top of the other.

The goal of good science is to know which small, out-of-place bits of fact are critical pieces in the puzzle, which tiny irregularities are harbingers of vast, undiscovered laws. As physicist Steven Weinberg says, "There is nothing in any single disagreement between theory and experiment that stands up and waves a flag and says, 'I am an important anomaly.'"

The differences between Einstein's laws and Newton's laws are truly very small—all but imperceptible, for that matter, except at speeds close to the velocity of light. You never notice that time slows down on a trip across the country, or that everyone in the plane gets slightly more massive during the flight. You could never notice the difference between gravity as a Newtonian force and as an Einsteinian curvature of space. Yet the significance of those small differences is enormous. As Richard Feynman points out, Einstein's laws make Newton's laws only a little bit wrong, but "philosophically, we are completely wrong," with Newton's laws. "This is a very peculiar thing about the philosophy, or the ideas, behind the laws. Even a very small effect sometimes requires profound changes in our ideas."

In the end, it was the small differences that Darwin noticed in species of turtles and birds and iguanas that led him down the path toward his evolutionary theory in the first place: the island-by-island variations in shell and beak and color. Even today, confidence in evolution is continually bolstered by such small irregularities as Stephen Jay Gould's now well-known "panda's thumb"—a thumb that is really an appendage jury-rigged out of an overgrown wristbone. Nearly perfect design is not good evidence for evolution, Gould argues, because it would more likely be the handiwork of a nearly perfect Creator. Rather, he says, "odd arrangements and funny solutions are the proof of evolution—paths that a sensible God would never tread but that a natural process, constrained by history, follows perforce."

Most of the time, small differences really *are* small differences. Once they're understood, you can relax about them. The occasional odd signal can serve as a chink in the armor of our complacency—a glitch in the works that jars us out of our accustomed ways of looking at things.

Big differences, when you think about it, are much more difficult to notice. You don't notice the motion of the earth even though it is spinning around its center at 1,000 miles per hour, and around the sun at almost 20 miles per second. You don't notice the flow of your own blood or the activities of your own cells. Major social and economic trends often pass us invisibly, because they build up so slowly. You can't feel the speed of your 747 even when it's whizzing along at 500 miles per hour.

Sometimes you need to hit an air pocket before you know that you're flying. Perhaps the biggest thing that small differences can do is open our eyes to the presence of larger, unexpected truths.

FORCES AND INFLUENCES

THE IDEAS in this book were molded largely from the works and thoughts of others—from the ideas and opinions of Albert Einstein to the writings of people like Guy Murchie and Lincoln Barnett and the always original insights of "my friend the physicist." Some of these people appear so often on the preceding pages that it seems only fitting to properly introduce them here:

The late FRANK OPPENHEIMER, director of The Exploratorium in San Francisco, became my friend the physicist when I first stumbled upon his amazing museum in 1972. I think that The Exploratorium resembles nothing so much as the inside of Frank's brain: a rather chaotic but deeply connected combination of art, science, philosophy, education, politics, and pure play. The Exploratorium is widely known as the world's best science museum, but in fact it is a museum of human perception. Since the beginning, a battered sign on the shop has announced: HERE IS BEING CREATED AN EXPLORATORIUM—A COMMUNITY MUSEUM DEDICATED TO HUMAN AWARENESS. Human awareness was always Frank's primary concern, since his days wandering New York City's water towers late at night and writing essays on the view from the top, to his student experiences at Johns Hopkins and Caltech practicing physics and the flute, through the years he worked on the atomic bomb project at Los Alamos, to his adventures as a cosmic ray physicist, to ten years of virtual political exile (the price he paid for his pacifist views) and cattle ranching in the Colorado Rockies, to his return to teaching, and finally the culmination of all his experiences in The Exploratorium. Frank received a great number of honors for his work, including two Guggenheim Fellowships and distinguished service awards from Caltech, the American Association of Physics Teachers, and the American Association of Museums.

VICTOR WEISSKOPF should really be called "my other friend the physicist." I first met him through his excellent book *Knowledge and Wonder,* which is the only book other than *The Exorcist* I have stayed up all night to read. Weisskopf has been Institute Professor emeritus at MIT, National Medal of Science winner, director of CERN during its most important formative years, president of the American Academy of Arts and Sciences, and member of the Pontifical Academy of Sciences, where he was deeply involved in issues of nuclear disarmament. In addition he has been deeply dedicated to what some scientists still scorn as "popularization"; he always tried to make time to work with science writers and explain the central ideas of physics to the general public. Weisskopf, like Oppenheimer, has always been remarkably and refreshingly relaxed about other people using his ideas. He has often said, "The only sin is if you hear a good idea and you don't use it."

PHILIP MORRISON, Institute Professor at MIT, astrophysicist, author, book reviewer and columnist for *Scientific American,* is probably as well known as a spellbinding lecturer and elegant writer as for his work in physics. While it is hard to pick favorites among his popular writings, the collection *Nothing Is Too Wonderful to Be True* probably best reflects his range and style. With his wife, Phylis, he produced the ever-enchanting classic *Powers of Ten.* Phil's approach is always original, and he is a master at turning the obvious inside out and forcing you to reevaluate your assumptions, whether the subject is atoms, disarmament or the search for extraterrestial life. On the many occasions we talked at MIT and at the Oppenheimers' house in San Francisco, he has left me both marveling at my own naïveté and filled to the brim with new ideas. Both Phil and Phylis contributed enormously to the development of The Exploratorium.

The late RICHARD FEYNMAN was once described as "the smartest man in the world." True or not, he was certainly one of the most colorful characters in the physics community. A Nobel Prize winner for his work in quantum electrodynamics, accomplished player of bongo drums, and creator of the famous Feynman diagrams for visually representing subatomic events, he was a theoretical physicist at Caltech, where his Queens accent was amusingly out of place. I once lucked into an hour-long talk with Feynman (he was notoriously press shy), but much of his material presented here is from the indispensable *Feynman Lectures on Physics* and also his book *The Character of Physical Law.*

The late GUY MURCHIE was a well-known science writer and author of *Song of the Sky, Music of the Spheres,* and *The Seven Mysteries of Life.* His books overflow with facts, quotes, anecdotes, and most of all enthusiasm for all things natural. I take great pride in having introduced Guy Murchie to The Exploratorium—and vice versa.

ALBERT EINSTEIN, of course, was the universally acclaimed genius who invented (or discovered, if you will) both the special theory of relativity ($E = mc^2$, time dilation, and all that) and the general theory of relativity (curved space, black holes, and all that). He revolutionized the way people think about time, space, matter, energy, motion, and other fundamental phenomena. But far more than a brilliant scientist, Einstein was a great humanist who talked, wrote, and worried about war, the human condition, tyranny, and most of all about the proliferation of nuclear bombs—something, he once said, that has changed everything but our way of thinking.

J. ROBERT OPPENHEIMER, Frank's older brother, is often credited with developing the first school of theoretical physics in the United States after his return from Europe in the 1920s. As the scientist in charge of Los Alamos, he is often known as father of the atomic bomb. Still, he was primarily known as a great teacher, deep thinker, and, in the end, something of a martyr for his political views: He was stripped of his security clearance during the McCarthy purges of the 1950s, partly because of his opposition to Edward Teller's determination to produce the vastly more powerful hydrogen bomb.

GEORGE GAMOW was an eccentric and important physicist who was one of the first major contributors to the now almost universally accepted Big Bang theory of the origins of the universe.[15] His enchanting stories of how a poor bank clerk came to learn about relativity and quantum physics (*Mr. Tompkins in Wonderland* and *Mr. Tompkins Explores the Atom*) are a delightful introduction for anyone interested in modern science. I also highly recommend his *One, Two, Three... Infinity, Biography of Physics,* and *Biography of the Earth.*

STEPHEN JAY GOULD is the outspoken and original Harvard biologist and geologist who writes those wonderful essays for *Natural History* magazine, collected in *Ever Since Darwin, The Panda's Thumb, Wonderful Life,* and many others. Gould rarely makes a point about

---

[15]Actually, what's universally accepted is rapid expansion of the universe in all directions—which implies that at one time in the distant past all the matter and energy was compressed together into one infinitesimally small point.

evolutionary biology without also drawing a parallel to broader aspects of human affairs.

VERA KISTIAKOWSKY was an experimental physicist and professor at MIT who took the time on several occasions to review my various writings and to talk with me about physics.

SIR JAMES JEANS was a British astronomer and physicist whose research covered a wide range from molecular physics to quantum theory and cosmology. He ceased research in 1928 (after he was knighted) to popularize science. His lectures and radio speeches were published in *The Universe Around Us* and *The Mysterious Universe.*

SIR ARTHUR EDDINGTON was a contemporary of Jeans's (the two disagreed on aspects of both astronomy and philosophy) who specialized in relativity theory. In fact, Eddington was the first to interpret Einstein's theory of relativity in plain English. Einstein considered Eddington's presentation the finest in any language.

NIELS BOHR was a Danish physicist who is widely known as the father of quantum mechanics: He inspired a whole generation of physicists with his ideas about science and its implications in human thought. Bohr was the first to attribute the specific properties of atoms to the fact that events within an atom (like the emission of light) happen only as a whole—a quantum leap, so to speak. He also developed the idea of complementary descriptions to reconcile this strangely particlelike aspect of radiation with its wavelike character.

ISAAC NEWTON was the seventeenth-century scientist who is pictured by most people as sitting under a tree waiting for an apple to fall on his head. True or apocryphal, it was Newton who first saw that the fall of the apple and the orbit (or "fall") of the moon were propelled by the same force—gravity. His three famous laws of motion (every action has an equal and opposite reaction, and so on) are familiar to every schoolchild. He created the calculus, first realized that white light was actually a mixture of the full spectrum of colors, and in the end was obsessed with alchemy and mysticism.

Aristotle's image of the universe as static, finite, and earth-centered dominated scientific thinking on and off for a millennium and a half. His primary contributions were not in physics, but he is often credited (or rather blamed) for getting future scientists off on the wrong tack—especially with his laws of motion, which erroneously assumed that the "natural" state of a body was at rest and that things would naturally come to a stop if they were no longer being pushed by a force.

Nicolaus Copernicus was the fifteenth-century astronomer who is credited with discovering that the earth moves around the sun and not vice versa. Many historical accounts say that the new Copernican system greatly simplified the old system developed by the Greek astronomer and mathematician Ptolemy, which described the motions of the planets, sun, and moon as a complicated collection of epicycles—circles within circles. But Copernicus's system also required many complicated motions, primarily because he saw the motions of heavenly bodies as perfect circles, even though the true orbits of the planets are ellipses.

Galileo Galilei was the sixteenth- and seventeenth-century scientist who is credited with bringing back the importance of experimental evidence into the study of physical phenomena. Galileo is often pictured standing at the top of the Leaning Tower of Pisa and dropping a feather and a rock to prove that they fall at the same rate (an impossibility, unless Pisa existed in a vacuum). He is also credited with noticing that the period of a pendulum depends only on the length of the pendulum and not on the size of the swing; for inventing telescopes; for discovering mountains on the moon, numerous new stars, and the moons of Jupiter—the first evidence that planets other than our own could have satellites.

Johannes Kepler was a contemporary of Galileo who spent his life searching for cosmic harmonies in the "music of the spheres." Probably his most important contribution was his recognition that the planets were held in orbit around the sun by a *force;* he also developed the three famous laws of planetary motion, which finally proved that the orbits of the planets were ellipses and not circles.

Of course, this is only a partial list, organized more or less by frequency of appearance. It would be all but impossible to list all of the forces and influences that shaped this book. Many more appear in the Selected Bibliography.

SELECTED BIBLIOGRAPHY AND RECOMMENDED READING

Isaac Asimov, *Asimov On Chemistry, Asimov On Physics*

Hans Christian von Baeyer, *The Fermi Solution: Essays on Science, Taming the Atom: The Emergence of the Visible Microworld*

Adolph Baker, *Modern Physics and Antiphysics*

Lincoln Barnett, *The Universe and Dr. Einstein*

Marcia Bartusiak, *Through a Universe Darkly*

Max Born, *The Natural Philosophy of Cause and Chance*

Jacob Bronowski, *The Ascent of Man, Science and Human Values*

Nigel Calder, *Einstein's Universe*

Hendrik Casimir, *Haphazard Reality: Half a Century of Science*

Margaret Cheney, *Tesla: Man out of Time*

Annie Dillard, *Pilgrim at Tinker Creek*

Sir Arthur Eddington, *Space, Time and Gravitation: An Outline of the General Relativity Theory*

Albert Einstein, *Essays in Physics, Ideas and Opinions*

Albert Einstein and Leopold Infeld, *The Evolution of Physics*

Loren Eiseley, *The Unexpected Universe*

Richard Feynman, *The Character of Physical Law, The Feynman Lectures on Physics*

A. P. French, editor, *Einstein: A Centenary Volume*

George Gamow, *Mr. Tompkins Explores the Atom, Mr. Tompkins in Paperback, Mr. Tompkins in Wonderland, Biography of Physics, Gravity*

Martin Gardner, *The Whys of a Philosophical Scrivener*

Larry Gonick and Art Huffman, *The Cartoon Guide to Physics*

Richard L. Gregory, *Eye and Brain: The Psychology of Seeing, The Intelligent Eye, Mind in Science: A History of Explanations in Psychology and Physics*

Paul G. Hewitt, *Conceptual Physics, Thinking Physics*

Sir James Jeans, *Physics and Philosophy*

Daniel Kevles, *The Physicists: The History of a Scientific Community in Modern America*

Arthur Koestler, *The Sleepwalkers: A History of Man's Changing Vision of the Universe*

Lawrence Krauss, *Beyond Star Trek: Physics from Alien Invasions to the End of Time, Fear of Physics: A Guide for the Perplexed, The Physics of Star Trek*

Robert March, *Physics for Poets*

P. B. Medawar, *Advice to a Young Scientist*

Philip Morrison, *Nothing Is Too Wonderful to Be True*

Philip Morrison and Phylis Morrison, *The Ring of Truth: An Inquiry into How We Know What We Do*

Guy Murchie, *Music of the Spheres: The Material Universe from Atom to Quasar, Simply Explained; The Seven Mysteries of Life*

J. Robert Oppenheimer, *Science and the Common Understanding*

Ilya Prigogine, *From Being to Becoming*

B. K. Ridley, *Time, Space and Things*

Carl Sagan, *Cosmos*

Lee Smolin, *The Life of the Cosmos*

Peter S. Stevens, *Patterns in Nature*

Kip Thorne, *Black Holes and Time Warps: Einstein's Outrageous Legacy*

James Trefil, *The Unexpected Vista*

Judith Wechsler, editor, *On Aesthetics in Science*

Victor Weisskopf, *Knowledge and Wonder: The Natural World as Man Knows It, Physics in the Twentieth Century*

Steven Weinberg, *Dreams of a Final Theory, The First Three Minutes: A Modern View of the Origin of the Universe*

Joseph Weizenbaum, *Computer Power and Human Reason: From Judgment to Calculation*

Frank Wilczek and Betsy Devine, *Longing for the Harmonies*

INDEX